Nutraceuticals - Past, Present and Future

Edited by María Chávarri Hueda

Published in London, United Kingdom

IntechOpen

Supporting open minds since 2005

Nutraceuticals - Past, Present and Future
http://dx.doi.org/10.5772/intechopen.78440
Edited by María Chávarri Hueda

Contributors
Charu Gupta, Dhan Prakash, Sevda Inan, Ali Hossein Khani, Seyed Milad Mousavi Jazayeri, Elahe Ebarhimi, Ayoub Farhadi, Cristina Barrera, Leidy Indira Hinestroza, Noelia Betoret, Lucía Seguí, Shantal Windvogel, María Chávarri Hueda

First published in London, United Kingdom, 2020 by IntechOpen
IntechOpen is the global imprint of INTECHOPEN LIMITED, registered in England and Wales, registration number: 11086078, 7th floor, 10 Lower Thames Street, London, EC3R 6AF, United Kingdom
Printed in Croatia

British Library Cataloguing-in-Publication Data
A catalogue record for this book is available from the British Library

Additional hard and PDF copies can be obtained from orders@intechopen.com

Nutraceuticals - Past, Present and Future
Edited by María Chávarri Hueda
p. cm.
Print ISBN 978-1-83881-161-7
Online ISBN 978-1-83881-162-4
eBook (PDF) ISBN 978-1-83881-163-1

For EU product safety concerns:
IN TECH d.o.o., Prolaz Marije Krucifikse Kozulić 3, 51000 Rijeka, Croatia, info@intechopen.com or visit our website at intechopen.com.

Meet the editor

Maria Chávarri Hueda received her MS in Biological Sciences from Universidad de Navarra, Spain, in 1997. She obtained a PhD in Nutrition and Food Science from the University of the Basque Country. Dr. Hueda has experience in biotechnology and food science, specifically bioactive molecules and functional activity, probiotics, and nutritional status. She worked on the "Influence of the lipid source of the diet on various aspects of hepatic metabolism of triglycerides and cholesterol". Over the last few decades, she has worked as a senior researcher in the area of food and health at Tecnalia Research & Innovation, Technological Development Center. She has focused her studies on bioactive molecules of food and plant origin and their functional activities. She also studies probiotics, with the aim of developing functional foods and nutraceuticals.

Contents

Preface

Society is increasingly aware that food can help prevent the development of certain diseases. This, together with the increase in life expectancy, is changing the trend of food consumption. For this reason, it is important to know which bioactive compounds possess a functional activity; that is, which ones produce a beneficial effect in the organism and can improve human health, as well as which ones can be used to develop nutraceuticals.

This book comprehensively reviews and compiles information on molecules with effect in health and that can be used to develop nutraceuticals. Organized into five chapters, the book covers knowledge of different active compounds in the prevention and treatment of prevalent diseases.

The scientists involved in the writing of this book were selected and invited because of their recognized expertise and important contributions in their respective fields. Their work made the publication of this book possible.

This book will be of help to scientists, doctors, pharmacists, chemists and other experts in a variety of disciplines, both academic and industrial.

I would like to thank my daughters Paula and Lucia, and my husband Alex for their patience and love. I extend my apologies for many hours spent on the editing of this book, which kept me away from them.

Dr. María Chávarri Hueda
Health and Food Area,
Health Division,
Tecnalia Research and Innovation,
Miñano (Alava), Spain

Chapter 1

Introductory Chapter: Nutraceuticals as an Alternative to Maintain a Healthy Lifestyle

María Chávarri

1. Introduction

The nutraceutical concept began from a survey conducted in the United Kingdom, Germany, and France, which concluded that consumers gave more importance to diet than to hereditary factors or to exercise to achieve good health. The term nutraceutical was defined by the nutritionist and pharmacist Stephen DeFelice, president of the Foundation of Innovation Medicine, in 1989. According to him, "a nutraceutical is defined as a food, or part of a food, that provides medical or health benefits, including prevention and/or treatment of a disease" [1, 2]. Specifically, nutraceuticals are formed from active compounds obtained from plant foods (such as phytocomplexes) or from foods of animal origin, which are concentrated and provided in the appropriate pharmaceutical form, and also have a pharmacological effect and nutritional value [3]. These can be used effectively to prevent and even cure some diseases when their safety is demonstrated, have a greater bioavailability, and have clinically proven health effects [4]. In the case of Health Canada, it defines the term nutraceutical as "a product prepared from food, but sold in the form of pills, powder or other medicinal forms, which are generally not associated with food" [5].

Before the appearance of nutraceutical concept, the consumption of fruits and vegetables was related in a preventive and protective way with a lower risk of suffering from chronic and degenerative diseases [6]. These discoveries were and are correlated with the diversity of plants and their richness in bioactive compounds, natural substances capable of modulating one or more metabolic processes, thus promoting health conditions [7, 8]. These substances, called phytochemicals, have allowed the discovery and the development of numerous medications especially for the treatment of diseases such as cancer, infectious diseases, cholesterolemia, and immunological disorders among others [9].

In the literature, other terms such as phytochemicals, herbs, spices, botanical medicines, dietary supplements, and secondary metabolites can be found, which can be mistaken for nutraceuticals [10]. In the case of dietary supplements, these are known as "concentrated sources of nutrients or other substances with a physiological or nutritional effect to complement the diet" [11]. Nutraceuticals can be found in various forms (isolated compounds, dietary supplements, or whole foods). Therefore, all nutraceuticals are not dietary supplements, and all dietary supplements are not nutraceuticals [10]. The difference between nutraceuticals and functional food is that the former are bioactive ingredients of natural origin and obtained from different food matrices, while the latter is considered any fresh or processed food that ensures a healthy effect and/or prevents diseases in addition

IntechOpen

to having a nutritional function [12]. Therefore, a food is functional when it has nutraceutical ingredients.

One of the examples of nutraceuticals is seaweed. The consumption of seaweed as traditional food and complementary medicine was recorded 10,000 years ago [13]. Traditionally consumed in many Asian countries such as Japan, China, the Philippines, Indonesia, Malaysia, South Korea, and North Korea for centuries, the culinary use of seaweed began in Japan and China. In the case of western countries, its main application has been as a gelling agent and colloid for the food, pharmaceutical, and cosmetic industry. However, recently, in western countries, such as the USA and Europe, among others, the frequent use of seaweed as part of food has begun, especially for its beneficial and functional properties [14]. Currently, the great recognition of the beneficial properties of algae has allowed this ingredient to be introduced as part of food and beverages, making it an important product for the food industry, additives for functional foods, feed supplements, etc. [15].

Algae are a good source of nutrients such as proteins, vitamins, minerals, and dietary fiber; in this regard, the kelp dietary fiber is particularly rich in soluble fraction. If algae are compared with terrestrial vegetables, there are more beneficial components for health, such as ω-3 fatty acids and bioactive molecules. Algae synthesize various secondary metabolites that have antioxidant, anti-inflammatory, anticancer, and antidiabetic activity. Therefore, algae can be considered a natural source of great interest since they contain compounds with numerous biological activities and can be used for the development of nutraceuticals. However, currently many industrial products based on seaweed have not been developed. This is due to the prolonged studies of nutritional intervention in humans required to demonstrate that algae are an excellent ingredient for commercial nutraceuticals [13].

Nutraceuticals can be classified in various ways; however, normally, the chemical nature, the availability of food, and the mechanism of action of the bioactive compound are taken into account [4]. **Figure 1** shows a diagram with the categories and subcategories. Compounds that are acquired from nature directly and are used without change in their form are considered traditional nutraceuticals. These nutraceuticals are divided into (a) phytochemicals, (b) probiotic bacteria, (c) enzymes, (d) chemical compounds, (e) nutrients, and (f) plants.

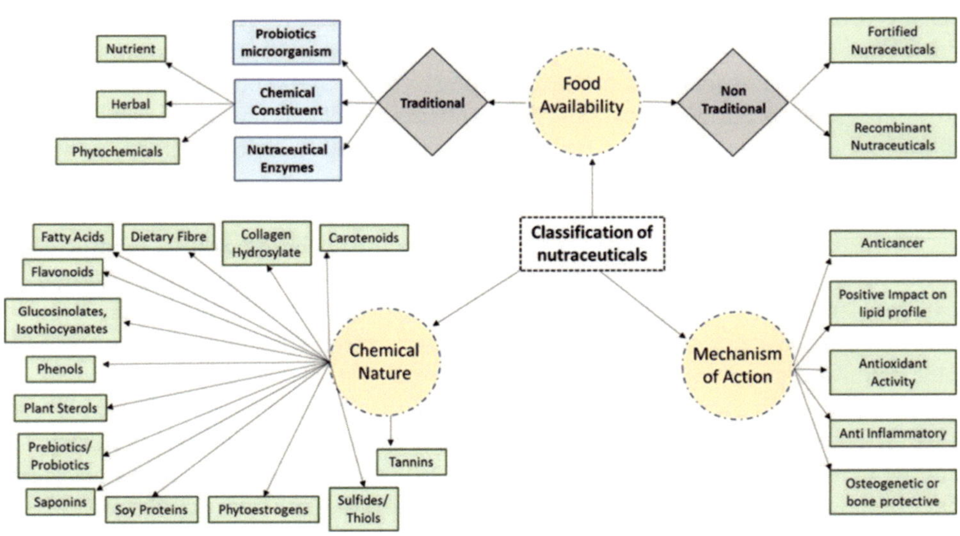

Figure 1.
Nutraceuticals classification [4].

Specifically, the nutraceutical market is a growing and relatively new section of the food industry. This growth has been due to several causes. On the one hand, consumers are increasingly concerned about their health and how it is achieved through food, and there is more and more evidence that relates to diet and health. In addition, more and more consumers want not to have to go to the doctor or not to consume medications; specifically, this fact is marked in the generation of Baby-Boomer. Considering the point of view of the consumers, nutraceuticals have several benefits [5]:

- The consumption of nutraceuticals will increase the health value of the individual's diet.
- An increase in the healthy value of the diet will help increase the lifespan of individuals.
- Nutraceuticals will help avoid particular medical deficiencies.
- They have beneficial psychological effects.
- Traditional medicines are more likely than nutraceuticals to generate harmful effects, so nutraceuticals are considered more natural.
- Nutraceuticals can be a very advantageous alternative for people with specific deficiencies or special needs.

Due above all to the increase in people's life expectancy, it is proposed to maintain a good health and well-being rather than having to treat diseases. Therefore, at present, the importance of preventing disease development is becoming clearer, being even more important than the treatment of diseases. The current progress in the treatment of diseases has been tremendous both in medicine and in drugs. At the same time, public health spending increases at the same rate as the progress of so-called diseases related to bad eating habits, including bad habits in people's lifestyle. In this sense, in recent years the term proactive medicine has appeared; focuses on the prevention of the development of diseases, reducing the cost of long-term chronic disease treatments, allowing an increase in well-being, improving the quality of life and state of people's health. Nutraceuticals are an alternative to help prevent the development of such diseases.

Author details

María Chávarri
Health and Food Area, Health Division, Tecnalia Research and Innovation, Miñano, Álava, Spain

*Address all correspondence to: maria.chavarri@tecnalia.com

References

[1] DeFelice SL. The Nutraceutical Revolution: Fueling a Powerful, New International Market. Como, Italy: Harvard University Advanced Management Program in Biomedical Research and Development; 1989

[2] DeFelice SL. The nutraceutical revolution: Its impact on food industry R&D. Trends in Food Science and Technology. 1995;**6**:59-61

[3] Maddi VS, Aragade PD, Digge VG, Nitalikar MN. Importance of nutraceuticals in health management. Pharmacognosy Reviews. 2007;**1**(2):377

[4] Ali A, Ahmad U, Akhtar J, Badruddeen, Khan MM. Engineered nano scale formulation strategies to augment efficiency of nutraceuticals. Journal of Functional Foods. 2019;**62**:103554

[5] Fernandes SD, Narayana RC, Narayanan AV. The emergence of India as a blossoming market for nutraceutical supplements: An overview. Trends in Food Science and Technology. 2019;**86**:579-585

[6] Boeing H, Bechthold A, Bub A, Ellinger S, Haller D, Kroke A, et al. Critical review: Vegetables and fruit in the prevention of chronic diseases. European Journal of Nutrition. 2012;**51**:637-663. DOI: 10.1007/s00394-012-0380-y

[7] Moiseeva EP, Almeida GM, Jones GD, Manson MM. Extended treatment with physiologic concentrations of dietary phytochemicals results in altered gene expression, reduced growth, and apoptosis of cancer cells. Molecular Cancer Therapeutics. 2007;**6**:3071-3079. DOI: 10.1158/1535-7163.MCT-07-0117

[8] Shanmugam MK, Kannaiyan R, Sethi G. Targeting cell signaling and apoptotic pathways by dietary agents: Role in the prevention and treatment of cancer. Nutrition and Cancer. 2011;**63**:161-173. DOI: 10.1080/01635581.2011.523502

[9] Thomford NE, Senthebane DA, Rowe A, Munro D, Seele P, Maroyi A, et al. Natural products for drug discovery in the 21st century: Innovations for novel drug discovery. International Journal of Molecular Sciences. 2018;**19**:E1578. DOI: 10.3390/ijms19061578

[10] Santana-GálvezJ,Cisneros-ZevallosL, Jacobo-Velázquez DA. A practical guide for designing effective nutraceutical combinations in the form of foods, beverages, and dietary supplements against chronic degenerative diseases. Trends in Food Science and Technology. 2019;**88**:179-193

[11] Carlsohn A, Cassel M, Linné K, Mayer F. How much is too much? A case report of nutritional supplement use of a high-performance athlete. The British Journal of Nutrition. 2011;**105**:1724-1728

[12] Shahidi F. Nutraceuticals and functional foods: Whole versus processed foods. Trends in Food Science and Technology. 2009;**20**:376-387. DOI: 10.1016/j.tifs.2008.08.004

[13] Ganesan AR, Tiwari U, Rajauria G. Seaweed nutraceuticals and their therapeutic role in disease prevention. Food Science and Human Wellness. 2019;**185**:1-12. DOI: 10.1016/j.fshw.2019.08.001

[14] Bocanegra A, Bastida S, Benedi J, Rodenas S, Sanchez-Muniz FJ. Characteristics and nutritional and cardiovascular-health properties of seaweeds. Journal of Medicinal Food. 2009;**12**(2):236-258

[15] Rajauria G, Cornish L, Ometto F, Msuya FE, Villa R. Identification and selection of algae for food, feed, and fuel applications. In: Seaweed Sustainability. UK: Academic Press; 2015. pp. 315-345

Chapter 2

The Potential Role of Nutraceuticals in Inflammation and Oxidative Stress

Sevda Inan

Abstract

Nutraceuticals are defined as a food or food ingredients that prevent and treat diseases. They contain dietary supplements like proteins, vitamins and minerals, compound derived from natural sources. They have functions about delaying, preventing and treating chronic inflammatory diseases due to the presence of the phytochemicals. They have anti-inflammatory effects by inhibiting of the activation of NF-κB, blocking the overexpression of tumor necrosis factor and interleukin-1, downregulation of the overexpression of cell adhesion molecules and inhibiting phospholipase A2, COX-2, lipoxygenease, iNOS, myeloperoxidase and inhibiting reactive oxygen species (ROS) generating enzyme activity and increasing ability to scavenge ROS. They have antioxidative role that can reduce the level of ROS and free radicals. They have effects on the process of lipid oxidation that inhibit or slow the formation of free alky radicals and cut off the free radical chain reactions.

Keywords: nutraceuticals, inflammation, oxidative stress, protective functions, disease

1. Introduction

In recent years, the consumption of natural products or functional foods are increased and enlarged segment of food industry. At the same time, Nutraceuticals are increased using as an alternative for pharmaceutical industry especially variety of diseases and cancers in humans and animals.

Firstly, nutraceutical is a term used by Stefane De Felice Nutraceuticals is defined as a food, bioactive products or food ingredients that prevent and treat diseases [1–4]. They are not drugs but they have pharmacologically active substance [2]. They contain dietary supplements like proteins, vitamins and minerals, compounds derived from natural sources. They provide health and medical benefits that delay, prevent and treat chronic inflammatory diseases due to the presence of the phytochemicals [1–4].

They have anti-inflammatory effects by inhibiting of the activation of NF-κB, blocking the overexpression of tumor necrosis factor and interleukin-1, downregulation of the overexpression of cell adhesion molecules and inhibiting phospholipase A2, COX-2, 5-LOX, iNOS, myeloperoxidase and inhibiting ROS generating enzyme activity and increasing ability to scavenge ROS. They have antioxidative role that can reduce the level of ROS and free radicals. They have effects on the process of lipid oxidation that inhibit or slow the formation of free alky radicals and cut off the free radical chain reactions. They have intracellular signaling pathway modular effects [2–4].

The foods including antioxidative nutraceuticals are fruits (grape, citrus, blueberries, strawberries, blackberries and crowberries), vegetables (tomato, beans, broccoli, beet, mushroom, corn, white cabbage, kale, cauliflower, spinach, garlic, onion, cacao beans and soybean), spices (rosemary, oregano and thyme), herbs (sage) and beverages (tea, wine) [4–6].

People's interest about nutraceuticals is increasing day by day due to various diseases. According to the global market data, China will be first nutraceutical market as lifestyle. The nutraceutical sector is affected by the stringent regulations and approval process of European Union. Due to the country, there are different names of laws on nutraceuticals. Nutraceuticals are using different definitions and terms including dietary supplement in USA, Natural Health Product in Canada, complementary medicines in Australia, food supplements in European Union and foods for special dietary in India [7].

With the increasing technology in the food, health and pharmaceutical sectors, the orientation to functional foods is increasing and the competition is accelerating. The sales of global market for nutraceuticals are expected to be US$250 billion by 2018 [8].

When nutraceuticals are evaluated by consumers, the consumption of food has undergone changes in the past three decades. The easing of access to media and internet, increasing in scientific studies and obesity related diseases are increased to sale nutraceutical products by consumers. Between 2018 and 2025 years, the growth rate of this sector is assumed to exceed 9.7%. The countries including Brazil, China, India, South Korea, Poland and Mexico are increasing to use functional foods. The global market of nutraceuticals is assumed to be $578.23 billion by 2025 at CAGR of 8.8% [9].

Inflammation is a protective response against the initial cause of cell injury. Inflammation is classified as acute and chronic. Acute inflammation is first response mechanism against infections, trauma, physical and chemical agents, which are induced wound healing. If this mechanism occurs persistent, it takes chronic phase [3, 10]. The process of inflammation contain vascular and cellular changes including of swollen, redness, local heating and loss of function. The permeability of capillaries is increased, exudate including the fluid and other elements leak into the body cavities. The inflammatory cells, leucocytes and other phagocytic cells migrate through the affected region. The lytic enzymes release from lysosomes of cells. During the inflammation, chemical mediators are synthesized proinflammatory cytokines (histamine, 5-hydroxytryptamine, bradykinin, leukotrienes and prostaglandins), selectins, integrins and immunoglobulins are stimulated for releasing [1]. Arachidonic acid metabolites including prostaglandins and leukotrienes are stimulated by the increasing expression of phospholipase A2. ROS are released from the inflammatory cells including neutrophils and macrophages. NADPH oxidase, xanthine oxidase and myeloperoxidase are seen increasing due to the ROS. The inflammatory cytokines, cell adhesion molecules and enzymes are regulated by the activation of the transcription factor NF-κB [3, 11].

ROS generate intracellularly as natural by endogenous and exogenous sources. Endogenous ROS including superoxide, hydrogen peroxide and nitric oxide (NO) have functions in cell signaling and homeostasis [12]. ROS has functions in regulation of cell survival. At the moderate levels of ROS signaling support cell proliferation and survival. At the upper levels of ROS cause cell death [12, 13]. There is a relationship between ROS production and oxidative stress that play a role on redox signaling from the organelle to the cytosol to nucleus [12, 14].

ROS are present in different cancer types and age related diseases as neurodegeneration, inflammation, diabetes, vision and sensory loss [12]. ROS and reactive nitrogen species damage significant biological molecules which are lipids, DNA, essential cellular proteins. Oxidative stress is imbalance between the formation of free radicals and antioxidant defense mechanism [15, 16].

Enzymatic and nonenzymatic antioxidant systems which are superoxide dismutase, catalase, glutathione peroxidase, lipid soluble vitamin E, carotenes and water soluble vitamin C arrange between ROS and antioxidants [4, 17, 18].

Oxidative stress starts the oxidation of polyunsaturated fatty acids (PUFA), proteins, DNA and sterols. The oxidative stress reduce in the body with consumption of fruits and vegetables including high amounts of anti-oxidative nutraceuticals and for this reason, incidence of cancer and cardiovascular diseases decrease [4, 6]. According to the recent studies, there is a relationship between ROS and atherosclerosis, vasospasm, cancers, trauma, stroke, asthma, hyperoxia, arthritis, heart attack, age pigments, dermatitis, cataractogenesis, retinal damage, hepatitis, liver injury and periodontitis [4, 19, 20].

2. Nutraceuticals

2.1 Vitamin E

Vitamin E (alpha-, beta-, gamma- and delta-tocopherol, alpha-, beta-, gamma- and delta-tocotrienol) is quite effective antioxidant and beneficial aspects for rheumatoid arthritis [4, 21, 22]. Also, vitamin E has anti-inflammatory effects in animal recent studies [4, 23]. Tocopherols and tocotrienols have nonpolar structures and consist in the lipid phase. Tocopherols are member of biological membranes and. Tocopherols have antioxidants property that defend polyunsaturated fatty acids into the membrane and LDL [4, 24]. The anti-inflammatory and anti-oxidant effects of Vitamin E and its derivatives are summarized in **Table 1**.

Vitamin E and derivatives	Anti-oxidant and anti-inflammatory effects	References
(Review literature study), (randomized, double-blind placebo-controlled human study, 400 mg for 3 months), (The transgenic KRN/NOD mice, 0.268 mg for 6 weeks)	Effects on rheumatoid arthritis against the inflammation and oxidative stress	Lee et al. [4], Aryaeian et al. [21], Bandt et al. [22]
(Review literature study), (30 and 500 ppm for 30 days in old mice)	Inhibition of cyclooxygenase activity in macrophages	Lee et al. [4], Beharka et al. [23]
(Review literature study in elderly cardiovascular patients)	Decreases risk of cardiovascular disease, anti-cancer activity and decreases incidence of Alzheimer's disease	Meydani [24]
(Review literature study)	Changes the level cholesterol and blocks oxidation of LDL.	Lee et al. [4]
(Review literature study)	Alterations of cell membrane integrity, cell division and cell signaling pathways. Stimulates indirectly prostaglandin and cytokines, directly stimulates T cell function. Reduces incidence of infectious diseases including respiratory infections and asthma	Lewis et al. [25]

Vitamin E and derivatives	Anti-oxidant and anti-inflammatory effects	References
(Different doses, review literature study)	Prevents and treats a multitude of age related diseases. Ameliorates of lipid profile and modulates suppression of the senescence- associated secretory phenotype	Malavolta et al. [26]
(The randomized clinical trials, ranging doses 33-800 IU)	Effects lonely cardiovascular diseases by reducing myocardial infarction	Loffredo et al. [27]
(The clinical review literature study)	Preventive and therapeutic functions in cardiovascular diseases.	Jain et al. [28]
(The consumption of different doses, review of literature study in human)	Prevents various types of cancer, heart disease and chronic ailments	Shahidi [29]
(The ranging doses between 500 IU/kg for 4 weeks in rats, 600 mg/kg in rats, 45 and 60 mg/kg in rats)	Anti-oxidant roles by decreasing the distribution of free radicals and modulating plasmatic lipoproteins in traumatic brain injury related dementia	Dobrovolny et al. [30]

Table 1.
The effects of Vitamin E and its derivatives, relevant to anti-inflammatory and anti-oxidant activity.

2.2 Carotenoids

They are classified as xanthophylls and carotenes. The carotenes have hydrocarbon and xanthophylls have oxygen [2, 4]. Carotenoids including alpha- carotene, lycopene, lutein, zeaxanthin, beta-carotene and beta-cryptoxanthin have antioxidant effects [1, 4]. The anti-inflammatory and anti-oxidant effects of carotenoids are summarized in **Table 2**.

Carotenoids	Anti-oxidant and anti-inflammatory effects	References
(The literature review study)	Functions on cell growth, embryonic development, vision property and immune system. Modulates activity of intracellular communication by interaction with nuclear receptors like pregnant X- receptor or retinoic acid receptor	Lushchak [2], Ruhl [31]
(The literature review study), (The prospective study of older women between 55-69 ages)	Protective roles against rheumatoid arthritis, atherosclerosis, cataracts, age-related muscular degeneration and multiple sclerosis	Al-Okbi [1], Lee et al. [4], Cerhan et al. [32]
(The ranging numbers and amounts of cases and exposure, the epidemiological review study), (The prospective cohort study between 1986 and 1992, in cases of 812 prostate cancer), (The review study related with the consumption of foods including different amounts of carotenoids)	Decreases the expansion of cervical, colon, prostate, rectal, stomach and other different of cancer types	Giovannucci [33], Giovannucci et al. [34], Giovannucci [35]
(The literature review study)	Blocks the formation of oxidized products of LDL cholesterol in coronary heart disease	Weisburger [36]

Carotenoids	Anti-oxidant and anti-inflammatory effects	References
(25 and 50 mg/kg of body sweight in mice for 3 days)	Antimutagenic effect	Polivkova et al. [37]
(Daily oral dose 10 mg/kg body weight and intraperitoneally 25 mg/kg body weight in female Wistar rats)	Neuroprotective activity	Sandhir et al. [38]
(Lycopene complex including 6% lycopene, 1.5% tocopherols, 1% phytoene and phytofluene, 0.2% beta-carotene for 10 days in rats at 6 mg/kg body weight)	Nephroprotective activity	Sahin et al. [39]
(The prospective randomized study in 159 primigravidas at the gestational time with the consumption of 2 mg oral lycopene daily for 77 women, placebo daily for 82 women)	Prevents preclampsia	Banerjee et al. [40]
(375 men and 576 women with hip fracture and nonvertebral fracture in elderly ages at different amounts of consumption of carotenoid and lycopene)	Decreases risk of hip fracture	Sahni et al. [41]
(The literature review study)	Anti-obesity functions by modulating insulin resistance and reducing blood glucose levels by regulation of cytokine expression from white adipose tissue	Gammone [42]
(In vitro research of 25 male Holstein calves in ages of 6–10 weeks and 3 Angus Heifers in ages of 8–30 weeks with doses of etinoic acid (1 μM) or β-carotene (8.3 μg/mL)	Promotes leukocyte apoptosis in bronchoalveolar lavage fluid and improves efferocytosis in macrophages	Duquette et al. [43]
(The review article study including animal and human in vitro researches)	Modulates intracellular signaling cascades, gene expression, and protein translation and blocks the translocation of nuclear factor κB to the nucleus. Inhibits Interleukin-8, prostaglandin E2 and oxidative strees damage by activating phase II and glutathione-S-transferases.	Kaulmann [44]
(The review article study)	Inhibits UV-induced cutaneous inflammation, pathologic keratinization, pigmentation and wrinkling	Imokawa [45]
(The intake of AIN-93G or AIN-93G + 10% Tangerine or red tomato powder for 35 weeks in mice)	Protects against the UVB-induced keratinocyte carcinoma.	Cooperstone et al. [46]
(The different amount of carotenoid content in commercial tomato hybrid Zebrino)	Cytoprotective functions by mitigating ROS production and protects against the glutatione depletion and lipid proxidation	Del-Giudice et al. [47]
(The randomized double-blinded clinical trial study in 51 patients with beta- carotene fortified symbiotic food including 0.05 g beta carotene)	Decreases levels of insulin, triglycerides, VLDL-cholesterol, total/HDL cholesterol ratio, plasma nitric oxide and glutathione	Asemi et al. [48]

Carotenoids	Anti-oxidant and anti-inflammatory effects	References
(The doses of 20 and 40 mg/kg xanthophylls in hens and chicks)	Decreases inflammatory mediators and apoptosis in chick tissues including liver, duodenum and jejenum	Gao et al. [49]
(The review article study)	Modulates macrophage polarization and stops the progression of non-alcoholic fatty liver disease and provides liver homeostasis	Ni et al. [50]
(The review article study)	Neuroprotective functions against the Alzherimer's disease that prevents progression this disease and modulates of Aβ peptide production and accumulation, oxidative stress and secretion of pro-inflammatory mediator	Mohammadzade h Honarvar [51]
(59 young participants with the supplementation of 13 and 27 mg/day macular carotenoids)	Reduces stress, cortisol and symptoms of emotional and physical health.	Stringham et al. [52]
(The carotenoid derivatives and crystalline lycopene from tomato extract)	Prevents cancer and protects bone health by inhibiting of the nuclear factor kappa B activity.	Linnewiel-Hermoni et al. [53]
(The different amounts of intake carotenoid in this review article study)	Reduces variety types of cancer including oral cavity and laryngeal regions.	Leoncini et al. [54]

Table 2.
The effects of Carotenoids, relevant to anti-inflammatory and anti-oxidant activity.

2.3 Phenolic compounds, polyphenols

The polyphenols are phenolic compounds that are defined as a benzene ring bearing one or more hydroxyl groups attached to the ring. They are including plants, vegetables, fruit, vines, tea, coffee and microalgae [2]. The phenolic compounds are classified as simple phenols, benzoquinoes, phenoic acids, acetophenones, phenylacetic acids, hydroxycinnamic acids, phenylpropens, coumarins, chromones, anthraquinones and flavonoids [4]. According to the recent articles, polyphenols have antioxidant [2, 4], anti-inflammatory [2, 55], anticancer [2, 56], antibacterial [2, 57], antiatherogenic [2, 58], antiangiogenic [2, 59], antimutagenic and free radical scavenging properties [4].

Flavonoids, which are water-soluble [3], are popular group of polyphenols and classified as flavones, flavonols, catechin or flavanols, anthocyanins and isoflavones. Flavonoids consist as free aglycones or with sugars connected the chemical structures to generate glycosides. Flavonoids have anti-inflammatory functions by inhibiting the enzymes responsible for production of phospholipase A2, cyclooxygenase and lipoxygenase [2, 3, 63]. The beneficial effects to health of phenolic compounds are listed below in **Table 3**.

2.3.1 Flavones

They are including apigenin, chrysin, baicalein, scutellarein and wogonin [2]. The anti-inflammatory and anti-oxidant effects of flavones are summarized in **Table 4**.

Polyphenols	Anti-oxidant and anti-inflammatory effects	References
(The review article study)	Anti-inflammatory effects	Lushchak et al. [2], Biesalski [55]
(The review article study)	Anti-cancer functions	Lushchak et al. [2], Fresco et al. [56]
(The review article study) (the commercial apple skin powder including 995.3 mg chlorogenic acid/100 g and 14.4 mg Trolox/g)	Anti-bacterial functions	Lushchak et al. [2], Du et al. [57]
(The interval of different age, dietary source of polyphenols and contents in this review article study)	Anti-atherogenic functions	Lushchak et al. [2], Rimbach et al. [58]
(The review article study)	Anti-angiogenic functions. Contributes formation of ROS by inhibiting enzymes or chelating trace elements	Lushchak et al. [2], Corradini et al. [59]
(The review article study)	Anti-mutagenic and free radical scavenging properties	Lee et al. [4]
(The serial review article study)	Modulates of intracellular communications in the phosphoinositide 3-kinase, Akt- protein kinase B, tyrosine kinase and protein kinase C signaling cascade	Lushchak et al. [2], Williams et al. [60]
(The rat kidney study with the different phenolic contents and amounts in foods)	Inhibits the angiotensin converting enzyme in cardiovascular system	Lushchak et al. [2], Actis-Goretta et al. [61]
(The review article study), (The 17 hypercholesterolemic male patients with the consumption of 40-90 g/day macadamia nuts for 4 weeks)	Decreases influence of inflammation, alters the gene expression of antioxidant enzymes and reduces the risk of cardiovascular disease and certain type of cancer	Al-Okbi [1], Lee et al. [4], Garg et al. [62]
(The review article study)	Anti-inflammatory functions	Chatterjee et al. [63]
(The review article study with different phenolic compounds activities and amounts)	Neuroprotective and anticonvulsive effects on brain tissue against the oxidative stress by binding to the benzodiazepine site on GABAA receptor	Diniz et al. [64]
(The randomized, controlled, double blind cross over human study, Olive oil including 80 mg phenolic compounds/kg, Olive oil including 500 mg phenolic compounds for 3 weeks)	Improves the proportions of IgA coated bacteria and plasma levels of C-reactive protein	Martin-Pelaez et al. [65]
(The review article study)	Regulates toll like receptor, inhibits cyclooxygenase, phospholipase A2 and anti-oxidant enzymes including xanthine oxidase	Yahfoufi et al. [66]

Table 3.
The effects of phenolic compounds, relevant to anti-inflammatory and anti-oxidant activity.

Flavones	Anti-oxidant and anti-inflammatory effects	References
(The review article study), (250 μM quercetin (specific activity, 52.9 mCi/mM) for 10 min by injecting of *Xenopus laevis* oocytes)	Beneficial effects including as GLUT inhibitors in diabetes	Lee et al. [4], Kwon et al. [67]
(The review article study)	Cyclooxygenase inhibitory ability in cancer	Lee et al. [4], Kinghorn et al. [68]
(The treatment with 2.5–20 μM apigenin in cell culture including human prostate cancer PC-3 and 22Rv1)	Anti-tumoral activity that inhibits of the p-IKKα, NF-κB/p65, cell proliferation, invasion of prostat cancer cells	Shukla et al. [69]
(The study including phosphorylating five flavones and showing pancreatic cholesterol esterase inhibitory functions by IC_{50})	Acts as pancreatic cholesterol esterase inhibitor	Lee et al. [4], Peng et al. [70]
(The review article study)	Reduces neurodegeneration	Lee et al. [4], Gasiorowski et al. [71]
(The review article study), (Apigenin which is isolated from Cordia dichotoma bark, is received 5 mg/kg, p.o. in Male Swiss mice)	Treats colitis and reduces inflammatory enzymes	Lee et al. [4], Ganjare et al. [72]
(The review article studies)	Decreases the expression of tumor necrosis factor alpha and interleukin-6 in macrophages cells	Chatterjee et al. [63], Wu and Schauss [73]
(The review article study)	Anti-arthritic functions	Laev et al. [74]
(The review article study in vitro and in vivo)	Anti-inflammatory functions for neurodegenerative disease	Nabavi et al. [75]
(The review article study about anti-oxidant, anti-cancer, anti-tumoral activity, anti-inflammatory and hepato-protective functions of dietary flavonoids)	Inhibits thromboxane synthesis in animal model and decreases iNOS and COX-2 expression	Xiao et al. [76]
(50 mg/kg of body weight doses of apigenin was injected intraperitoneally in Male C57BL/6J mice)	Immunomodulatory effects that reduces NF-κB activity in the lungs and inhibits leukocyte infiltration	Cardenas et al. [77]
(1, 10, 25, 50, 75 and 100 μM concentrations of flavonoids were added Murine C2C12 cell culture medium)	Protective effects on lipopolysaccharide related muscle atrophy	Shiota et al. [78]

Table 4.
The effects of flavones, relevant to anti-inflammatory and anti-oxidant activity.

2.3.2 Flavonols

They are protective functions from UV radiation [2]. They are including kaempferol, quercetin, myricetin, galangin and morin [2]. They have beneficial effects on different conditions and diseases related oxidative stress and inflammation. These effects are summarized in **Table 5**.

Flavonols	Anti-oxidant and anti-inflammatory effects	References
(At flavonoid concentrations of 10–70 μmol/L were applied in HL-60, U937 and Jurkat cells)	Inhibits intracellular accumulation of ascorbic acid	Park et al. [79]
(The ranging content amounts of oolong tea leaves including 54 polyphenols were evaluated on the pancreatic lipase activity in vitro)	Inhibitory functions on pancreatic lipase with diabetes mellitus	Nakai et al. [80]
(At the concentrations of 0, 50, 100 and 250 μM of quercetin for 24–72 hours were added cell culture medium including mouse embryo 3T3-L1 cells)	Anti-oxidant activity on cell apoptosis	Hsu et al. [81]
(The oral treatment doses at 2.8 g/kg in male Wistar rats, aged 8–10 weeks)	Anti-oxidant and renoprotective effects in streptozotocin-diabetic rats	Liu et al. [82]
(The extracts of Ficus carica Linn. (Moraceae) leaves and fruits and Morus alba Linn. root barks (Moraceae) were given 50 and 150 mg/kg in adult female Swiss albino rats)	Hepatoptotective effects on tetrachloride-related oxidative stress and injury in rat liver tissue	Singab et al. [83]
(At the dose of 5 μM kaempferol was added cell culture medium including rat osteoblast-like UMR106 cells)	Regulates bone sialoprotein gene transcription and new bone formation	Yang et al. [84]
(At the doses of 50 and 100 mg/kg kaempferol were given orally in animal model study)	Regulates cyclooxygenase, inhibits production of nitric oxide	Mahat et al. [85]
(The review article study), (The double-blind study was given orally 4 x 500 mg quercetin in non-smoking, un-treated sarcoidosis individuals)	Decreases of oxidative stress and inflammation in sarcoidosis, colonic damage and allergic airway conditions	Chatterjee et al. [63], Boots et al. [86]
(The mice were fed Western diet including 0.05% quercetin for 18 weeks)	Modulates on accumulation and activation of immune cells and increases expression of mitochondrial gene in adipose tissue.	Kobori et al. [87]
(At the doses of 10 and 50 mg/kg quercetin were given intraperitoneally to the male Sprague Dawley rats)	Neuroprotective and anti-oxidant effects in subarachnoid hemorrhage, inhibits brain damage and edema.	Dong et al. [88]
(At the doses of 0 and 210 μM quercetin or taraxasterol were added cell culture medium including human umbilical vein endothelial cells)	Anti-atherosclerotic and cardioprotective effects against the oxidative stress and inflammation	Yang et al. [89]

Table 5.
The effects of flavones, relevant to anti-inflammatory and anti-oxidant activity.

2.3.3 Flavanones

They have important effects that regulate on the inflammatory process and oxidative stress. These beneficial effects are summarized in **Table 6**.

2.3.4 Catechin or flavanols

They are found in variety of fruits (apples, apricots, blackberries and grapes), red wine, black tea and cocoa [2]. For example; the long-term consumption of tea

Flavanones	Anti-oxidant and anti-inflammatory effects	References
(The review article study)	Inhibitory effects on carcinogenesis	Kinghorn et al. [68]
(At the concentrations of 5 and 25 μM of chalcones and flavanones were given in the vitro study)	Regulate LDL oxidation in atherosclerosis	Miranda et al. [90]
(The review article study)	Have functions in anti-malarial chemotherapy	Kumar et al. [91]
(The review article study)	Anti-inflammatory effects	Kontogiorgis et al. [92]
(The review article study)	Anti-angiogenic effects	Mojzis et al. [93]
(At the treatment daily doses of 100 mg/kg naringenin were applied to the female BALB/c mice)	Decrease lung metastases in a breast cancer model	Qin et al. [94]
(At the dose of 50 mg/kg naringenin was applied to the adult male albino rats)	Anti-oxidative stress related hepatic damage in rats	Prabu et al. [95]
(At the concentration of 0.25 mmol of naringin derivatives was applied by agar dilution technique and direct contact assaying)	Anti-bacterial roles in pathogenic strains	Celiz et al. [96]
(At the doses of 50 mg/kg of quercetin and naringenin were applied to mice intraperitoneally)	Protect DNA in alloxan-induced diabetic mice	Orsolic et al. [97]
(At the doses of 30–200 μmol/L were treated into the cell culture including macrophage cell line RAW 274.6 and BV2 microglia)	Inhibit synthesis of nitric oxide and expression of cyclooxygenase-2 in macrophages and microglia	Chao et al. [98]

Table 6.
The effects of flavanones, relevant to anti-inflammatory and anti-oxidant activity.

inhibits low grade inflammation [73]. The chronic consumption of dark chocolate reduces serum C-reactive protein concentrations in blood circulation [63, 73]. The other effects on inflammatory and oxidative stress are summarized in **Table 7**.

Catechin	Anti-oxidant and anti-inflammatory effects	References
(The review article including clinical and experimantal studies)	Cardioprotective effects by inhibiting the NF-κB initiated production of cytokines and adhesion molecule	Bhardwaj et al. [99]
(The review article study) (at the mixture of cathechin, caffeic acid and resveratrol doses of 40 and 160 mg/kg body weight/day were given to the apoE KO mice for 8 weeks)	Anti-inflammatory and anti-atherogenic functions	Wu and Schauss [73], Norata et al. [100]
(The dentifrice including 1.0% green tea catechin was applied to the male Wistar rats)	Reduces gingival oxidative stress and periodontal inflammation	Maruyama et al. [101]
(After the massive hepatectomy, green tea extract catechins were applied to the male Wistar rats)	Anti-oxidative and anti-inflammatory effects on liver dysfunction with massive hepatectomy	Saito et al. [102]

Catechin	Anti-oxidant and anti-inflammatory effects	References
(The review article of clinical studies)	Prevents vascular problems related diabetes mellitus	Howes and Simmonds [103]
(At the double-blind randomized study was applied daily 200 mg flavanols for 28 male smokers)	Anti-inflammatory effects by decreasing expression of inflammatory genes in leukocytes and increases vascular health	Weseler et al. [104]
(At the randomized, placebo-controlled, double blind, crossover study was applied capsules including 1 g total catechin for 19 healthy men)	Cardioprotective effects by decreasing oxidation of low density lipoprotein and incidence of atherosclerosis	Suzuki-Sugihara et al. [105]
(The consumption of high flavanols chocolate for 4 weeks was applied to the overweight men between 45 and 70 ages)	Effects on endothelium related vasodilation. Increases leukocyte adhesion factors and vascular function	Esser et al. [106]

Table 7.
The effects of catechin or flavanols, relevant to anti-inflammatory and anti-oxidant activity.

3. Conclusion

Nutraceuticals are alternative or functional foods or ingredients that prevent or treatment of inflammatory and oxidative stress induced diseases. Nutraceuticals are cheaper and easier availability than prescription drugs. For this reason, consumer's demand has increased in recent years.

The effecting on pathogenesis and activity of diseases are also essential scientific subject for animal and human health. When the effects of nutraceuticals on oxidative stress and inflammatory related disease are discovered, usages of nutraceuticals in Pharmacology and scientific studies are seen huge growth. The relation between beneficial effects of nutraceuticals and diseases are required to research long-term multidisciplinary studies.

People are searching minimally processed food and want to benefit nutritional values and live healthy. For this reason nutraceutical market is growing day by day.

The aging, fast rising population, changing lifestyle and lifestyle induced diseases, healthcare research, increasing cases of cancer, economic and public problems are directed people to benefit better choices.

As a conclusion, nutraceuticals are important for nutrition of human and animal. The consumption of nutraceuticals is necessary to reduce effects of the oxidative stress and inflammation related diseases.

Conflict of interest

There is no conflict of interest for this chapter.

Author details

Sevda Inan
Department of Pathology, Faculty of Veterinary Medicine, Tekirdag Namik Kemal University, Tekirdag, Turkey

*Address all correspondence to: sevdainan@nku.edu.tr

References

[1] Al-Okbi SY. Nutraceuticals of anti-inflammatory activity as complementary therapy for rheumatoid arthritis. Toxicology and Industrial Health. 2012:1-12. DOI: 10.1177/0748233712462468

[2] Lushchak V. Microalgae of the Chlorophyceae class: Potential nutraceuticals reducing oxidative stress intensity and cellular damage. In: Vanessa BV, Rocio OB, Ruth RS, Paola TM, Adelaida HG, Egdar CE, editors. Oxidative Stress and Diseases, Intech Open; 2012. pp. 581-610. DOI: 10.5772/2535

[3] Watson RR, Preedy VR. Fruits, Vegetables, and Herbs. Bioactive foods in health promotion. In: Barve KH, Kulkarni YA, Gaikwad AB, editors. Nutraceuticals as Therapeutic Agents for Inflammation. 1st ed. Academic Press, Elsevier; 2016. pp. 121-147. DOI: 10.1016/B978-0-12-802972-5.00007-X

[4] Lee J, Koo N, Min DB. Reactive oxygen species, aging, and antioxidative nutraceuticals. Comprehensive Reviews in Food Science and Food Safety. 2004; **3**:21-33. DOI: 10.1111/j.1541-4337.2004.tb00058.x

[5] Bravo L. Polyhenols: Chemistry, dietary sources, metabolism, and nutrional significance. Nutrition Reviews. 1998;**56**:317-333. DOI: 10.1111/j.1753-4887.1998.tb01670.x

[6] Kaur C, Kapoor HC. Antioxidants in fruits and vegetables—The millennium's health. International Journal of Food Science and Technology. 2001;**36**:703-725. DOI: 10.1046/j.1365-2621.2001.00513.x

[7] Ganesh GNK, Ramachandran A, Suresh KR, Senthil V, Baviya Priyadharshini R. Nutraceuticals—A regulatory review. International Journal of Drug Regulatory Affairs. 2015;**3**(2):22-29. DOI: 10.22270/ijdra.v3i2.165

[8] Stirling C, Kruh W. Nutraceuticals: The future of intelligent food. Where food and pharmaceuticals converge. KPMG: Cutting Through Complexity. 2015. Available from: https://home.kpmg.com/content/dam/kpmg/pdf/2015/05/neutraceuticals-the-future-of-intelligent-food.pdf [Accessed: 2018-12-09]

[9] Jajodia S, Rawat DS, Sharma R. Indıan Nutraceuticals Market. Current Scenario & Future Trends. Assocham Indıa; 2018. Available from: https://mrssindia.com/wp-content/uploads/2018/07/Nutraceuticals-Report-Final.pdf [Accessed: 2018-12-09]

[10] Kumar V, Abbas AK, Aster JC. Inflammation and Repair Robbins Basic Pathology. 10th ed. Philadelphia: Elsevier; 2018. pp. 57-59 ISBN: 978-0-323-35317-5

[11] Huang MT, Ghai G, Ho CT. Inflammatory process and molecular targets for anti-inflammatory nutraceuticals. Comprehensive Reviews in Food Science and Food Safety. 2004;**3**:127-139. DOI: 10.1111/j.1541-4337.2004.tb00063.x

[12] Kador PF. Topical applied nutraceutical antioxidant formulation reduces ocular oxidative stress. Functional Foods in Health & Disease. 2017;7(1):17-35. DOI: 10.31989/ffhd.v7i1.297

[13] Trachootham D, Weigin L, Ogasawara MA, Del Valle NR, Huang P. Redox regulation of cell survival. Antioxidants & Redox Signaling. 2008; **10**(8):1343-1374. DOI: 10.1089/ars.2007.1957

[14] Murphy MP. How mitochondria produce reactive oxygen species. The Biochemical Journal. 2009;**417**(1):1-13. DOI: 10.1042/BJ20081386

[15] Gupta RC. Nutraceuticals. In: Milatovic D, Zaja-Milatovic S, Gupta RC, editors. Oxidative Stress and Excitotoxicity: Antioxidants from Nutraceuticals. Academic Press; 2016. pp. 401-413. DOI: 10.1016/B978-0-12-802147-7.00029-2

[16] Houston M. The role of nutrition and nutraceutical supplements in the treatment of hypertension. World Journal of Cardiology. 2014;**6**(2):38-66. DOI: 10.4330/wjc.v6.i2.38

[17] Thomas MJ. The role of free radicals and antioxidants: How do we know that they are working? Critical Reviews in Food Science and Nutrition. 1995;**2-35** (1–2):21-39. DOI: 10.1080/10408399509527683

[18] Wickens AP. Ageing and the free radical theory. Respiration Physiology. 2001;**128**(3):379-391. DOI: 10.1016/S0034-5687(01)00313-9

[19] Cohen JH, Kristal AR, Stanford JL. Fruit and vegetable intakes and prostate cancer risk. Journal of the National Cancer Institute. 2000;**92**(1):61-68. DOI: 10.1093/jnci/92.1.61

[20] Packer L, Weber SU. The role of vitamin E in the emerging field of nutraceuticals. In: Kramer K, Hoppe PP, Packer L, editors. Nutraceuticals in Health and Disease Prevention. New York: Marcel Dekker; 2001. pp. 27-43. DOI: 10.1093/ajcn/75.4.783

[21] Aryaeian N, Shahram F, Djalali M, Eshragian MR, Djazayeri A, Sarrafnejad A, et al. Effect of conjugated linoleic acids, vitamin E and their combination on the clinical outcome of Iranian adults with active rheumatoid arthritis. International Journal of Rheumatic Diseases. 2009;**12**(1):20-28. DOI: 10.1111/j.1756-185X.2009.01374.x

[22] Bandt MD, Grossin M, Driss F, Pincemail J, Babin-Chevaye C, Pasquier C. Vitamin E uncouples joint destruction and clinical inflammation in a transgenic mouse model of rheumatoid arthritis. Artritis & Rheumatology. 2002;**46**(2):522-532. DOI: 10.1002/art.10085

[23] Beharka AA, Wu D, Serafini M, Meydani SN. Mechanism of vitamin E inhibition of cyclooxygenase activity in macrophages from old mice: Role of peroxynitrite. Free Radical Biology and Medicine. 2002;**32**(6):503-511. DOI: 10.1016/S0891-5849(01)00817-6

[24] Meydani M. Effect of functional food ingredients: Vitamin E modulation of cardiovascular diseases and immune status in the elderly. American Journal of Clinical Nutrition. 2000;**71**(6): 1665S-1668S. DOI: 10.1093/ajcn/71.6.1665

[25] Lewis ED, Meydani SN, Wu D. Regulatory role of vitamin E in the immune system and inflammation. IUBMB Life. 2018:1-8. DOI: 10.1002/iub.1976

[26] Malavolta M, Pierpaoli E, Giacconi R, Basso A, Cardelli M, Piacenza F, et al. Anti-inflammatory activity of tocotrienols in age-related pathologies: A SASPected involvement of cellular senescence. Biological Procedures Online. 2018;**20**:22. DOI: 10.1186/s12575-018-0087-4

[27] Loffredo L, Perri L, Di Castelnuovo A, Iacoviello L, De Gaetano G, Violi F. Supplementation with vitamin E alone is associated with reduced myocardial infarction: A meta-analysis. Nutrition, Metabolism, and Cardiovascular Diseases. 2015;**25**(4):354-363. DOI: 10.1016/j.numecd.2015.01.008

[28] Jain AK, Mehra NK, Swarnakar NK. Role of antioxidants for the treatment of Nutraceuticals - Past, Present and Future cardiovascular diseases: Challenges and opportunities. Current

Pharmaceutical Design. 2015;**21**(30): 4441-4455. DOI: 10.2174/1381612821 666150803151758

[29] Shahidi F, De Camargo AC. Tocopherols and tocotrienols in common and emerging dietary sources: Occurrence, applications, and health benefits. International Journal of Molecular Sciences. 2016;**17**(10). DOI: 10.3390/ijms17101745

[30] Dobrovolny J, Smrcka M, Bienertova-Vasku J. Therapeutic potential of vitamin E and its derivatives in traumatic brain injury-associated dementia. Neurological Sciences. 2018; **39**(6):989-998. DOI: 10.1007/s10072-018-3398-y

[31] Ruhl R. Induction of PXR-mediated metabolism by beta-carotene. Biochimica et Biyophysica Acta. 2005; **1740**(2):162-169. DOI: 10.1016/j.bbadis.2004.11.013

[32] Cerhan JR, Saag KG, Merlino LA, Mikuls TR, Criswell LA. Antioxidant micronutrients and risk of rheumatoid arthritis in a cohort of older women. American Journal of Epidemiology. 2003;**157**(4):345-354. DOI: 10.1093/aje/kwf205

[33] Giovannucci E. Tomatoes, tomato-based products, lycopene, and Cancer: Review of the epidemiologic literature. Journal of the National Cancer Institute.1999;**91**(4):317-331. DOI: 10.1093/jnci/91.4.317

[34] Giovannucci E, Ascherio A, Rimm EB, Stampfer MJ, Colditz GA, Willett WC. Intake of carotenoids and retino in relation to risk of prostate cancer. Journal of the National Cancer Institute. 1995;**87**(23):1767-1776. DOI: 10.1093/jnci/87.23.1767

[35] Giovannucci E, Clinton SK. Tomatoes, lycopene, and prostate cancer. Experimental Biology and Medicine. 1998;**218**:129-139. DOI: 10.3181/00379727-218-44277

[36] Weisburger JH. Mechanisms of action of antioxidants as exemplified in vegetables, tomatoes and tea. Food and Chemical Toxicology. 1999;**37**(9–10): 943-948. DOI: 10.1016/S0278-6915(99) 00086-1

[37] Polivkova Z, Smerak P, Demova H, Houska M. Antimutagenic effects of lycopene and tomato Purée. Journal of Medicinal Food. 2010;**13**(6):1443-1450. DOI: 10.1089/jmf.2009.0277

[38] Sandhir R, Mehrotra A, Kamboj SS. Lycopene prevents 3-nitropropionic acid-induced mitochondrial oxidative stress and dysfunctions in nervous system. Neurochemistry International. 2010;**57**(5):579-587. DOI: 10.1016/j.neuint.2010.07.005

[39] Sahin K, Tuzcu M, SAhin N, Ali S, Kucuk O. Nrf2/HO-1 signaling pathway may be the prime target for chemoprevention of cisplatin-induced nephrotoxicity by lycopene. Food and Chemical Toxicology. 2010;**48**(10): 2670-2674. DOI: 10.1016/j.fct.2010.06.038

[40] Banerjee S, Jeyaseelan S, Guleria R. Trial of lycopene to prevent pre-eclampsia in healthy primigravidas: Results show some adverse effects. The Journal of Obstetrics and Gynecology Research. 2009;**35**(3):477-482. DOI: 10.1111/j.1447-0756.2008.00983.x

[41] Sahni S, Hannan MT, Blumberg J, Cupples LA, Kiel DP, Tucker KL. Protective effect of total carotenoid and lycopene intake on the risk of hip fracture: A 17-year follow-up from the Framingham osteoporosis study. Journal of Bone and Mineral Research: The Official Journal of the American Society for Bone and Mineral Research. 2009; **24**(6):1086-1094. DOI: 10.1359/jbmr.090102

[42] Gammone MA, D'Orazio N. Anti-obesity activity of the marine carotenoid fucoxanthin. Marine Drugs.

2015;**13**(4):2196-2214. DOI: 10.3390/md13042196.

[43] Duquette SC, Fischer CD, Feener TD, Muench GP, Morck DW, Barreda DR, et al. Anti-inflammatory effects of retinoids and carotenoid derivatives on caspase-3-dependent apoptosis and efferocytosis of bovine neutrophils. American Journal of Veterinary Research. 2014;**75**(12):1064-1075. DOI: 10.2460/ajvr.75.12.1064

[44] Kaulmann A, Bohn T. Carotenoids, inflammation, and oxidative stress—Implications of cellular signaling pathways and relation to chronic disease prevention. Nutrition Research. 2014; **34**(11):907-929. DOI: 10.1016/j.nutres.2014.07.010.

[45] Imokawa G. Intracellular signaling mechanisms involved in the biological effects of the xanthophyll carotenoid astaxanthin to prevent the photo-aging of the skin in a reactive oxygen species depletion-independent manner: The key role of mitogen and stress-activated protein kinase 1. Photochemistry and Photobiology. 2018. DOI: 10.1111/php.13034

[46] Cooperstone JL, Tober KL, Riedl KM, Teegarden MD, Cichon MJ, Francis DM, et al. Tomatoes protect against development of UV-induced keratinocyte carcinoma via metabolomic alterations. Scientific Reports. 2017;**7**(1):5106. DOI: 10.1038/s41598-017-05568-7

[47] Del Giudice R, Petruk G, Raiola A, Barone A, Monti DM, Rigano MM. Carotenoids in fresh and processed tomato (*Solanum lycopersicum*) fruits protect cells from oxidative stress injury. Journal of the Science of Food and Agriculture. 2017;**97**(5):1616-1623. DOI: 10.1002/jsfa.7910

[48] Asemi Z, Alizadeh SA, Ahmad K, Goli M, Esmaillzadeh A. Effects of beta-carotene fortified synbiotic food on metabolic control of patients with type 2 diabetes mellitus: A double-blind randomized cross-over controlled clinical trial. Clinical Nutrition (Edinburgh, Scotland). 2016;**35**(4):819-825. DOI: 10.1016/j.clnu.2015.07.009

[49] Gao YY, Jin L, Ji J, Sun BL, Xu LH, Wang QX, et al. Xanthophyll supplementation reduced inflammatory mediators and apoptosis in hens and chicks. Journal of Animal Science. 2016; **94**(5):2014-2023. DOI: 10.2527/jas.2015-9628

[50] Ni Y, Zhuge F, Nagashimada M, Ota T. Novel action of carotenoids on non-alcoholic fatty liver disease: Macrophage polarization and liver homeostasis. Nutrients. 2016;**8**(7). DOI: 10.3390/nu8070391

[51] Mohammadzadeh Honarvar N, Saedisomeolia A, Abdolahi M, Shayeganrad A, Taheri Sangsari G, Hassanzadeh Rad B, et al. Molecular anti-inflammatory mechanisms of retinoids and carotenoids in Alzheimer's disease: A review of current evidence. Journal of Molecular Neuroscience. 2017;**61**(3):289-304. DOI: 10.1007/s12031-016-0857-x

[52] Stringham NT, Holmes PV, Stringham JM. Supplementation with macular carotenoids reduces psychological stress, serum cortisol, and sub-optimal symptoms of physical and emotional health in young adults. Nutritional Neuroscience. 2018;**21**(4): 286-296. DOI: 10.1080/1028415X.2017.1286445

[53] Linnewiel-Hermoni K, Motro Y, Miller Y, Levy J, Sharoni Y. Carotenoid derivatives inhibit nuclear factor kappa B activity in bone and cancer cells by targeting key thiol groups. Free Radical Biology & Medicine. 2014;**75**:105-120. DOI: 10.1016/j.freeradbiomed.2014.07.024

[54] Leoncini E, Nedovic D, Panic N, Pastorino R, Edefonti V, Boccia S.

Carotenoid intake from natural sources and head and neck cancer: A systematic review and meta-analysis of epidemiological studies. Cancer Epidemiology, Biomarkers & Prevention. 2015;**24**(7):1003-1011 DOI: 10.1158/1055-9965.EPI-15-0053

[55] Biesalski HK. Polyphenols and inflammation: Basic interactions. Current Opinion in Clinical Nutrition and Metabolic Care. 2007;**10**(6):724-728 DOI: 10.1097/MCO.0b013e3282f0cef2

[56] Fresco P, Borges F, Marques MPM, Diniz C. The anticancer properties of dietary polyphenols and its relation with apoptosis. 2010;**16**(1):114-134. DOI: 10.2174/138161210789941856

[57] Du WX, Olsen CW, Avena-Bustillos RJ, Friedman M, McHugh TH. Physical and antibacterial properties of edible films formulated with apple skin polyphenols. Journal of Food Science. 2011;**76**(2):M149-M155. DOI: 10.1111/j.1750-3841.2010.02012.x

[58] Rimbach G, Melchin M, Moehring J, Wagner AE. Polyphenols from cocoa and vascular health—A critical review. International Journal of Molecular Sciences. 2009;**10**(10):4290-4309. DOI: 10.3390/ijms10104290

[59] Corradini E, Foglia P, Giansanti P, Gubbiotti R, Samperi R, Lagana A. Flavonoids: Chemical properties and analytical methodologies of identification and quantitation in foods and plants. Natural Product Research. 2011;**25**(5):469-495. DOI: 10.1080/14786419.2010.482054

[60] Williams RJ, Spencer JP, Rice-Evans C. Flavonoids: Antioxidants or signalling molecules? Free Radical Biology & Medicine. 2004;**36**(7):838-849. DOI: 10.1016/j.freeradbiomed.2004.01.001

[61] Actis-Goretta L, Ottaviani JI, Fraga CG. Inhibition of angiotensin converting enzyme activity by flavanol-rich foods. Journal of Agricultural and Food Chemistry. 2006;**54**(1):229-234. DOI: 10.1021/jf052263o

[62] Garg ML, Blake RJ, Wills RB, Clayton EH. Macadamia nut consumption modulates favourably risk factors for coronary artery disease in hypercholesterolemic subjects. Lipids. 2007;**42**(6):583-587. DOI: 10.1007/s11745-007-3042-8

[63] Chatterjee S, Jungraithmayr W, Bagchi D. Immunity and inflammation in health and disease. Emerging roles of nutraceuticals and functional foods in immune support. In: MJR H, editor. Phytochemicals as Anti-Inflammatory Nutraceuticals and Phytopharmaceuticals. Academic Press; 2018. pp. 363-388. DOI: 10.1016/B978-0-12-805417-8.00028-7

[64] Diniz TC, Silva JC, de Lima-Saraiva SR, Ribeiro FP, Pacheco AG, de Freitas RM, et al. The role of flavonoids on oxidative stress in epilepsy. Oxidative Medicine and Cellular Longevity. 2015;**2015**:171756. DOI: 10.1155/2015/171756

[65] Martín-Peláez S, Castañer O, Solà R, Motilva MJ, Castell M, Pérez-Cano FJ, et al. Influence of phenol-enriched olive oils on human intestinal immune function. Nutrients. 2016;**8**(4):213. DOI: 10.3390/nu8040213

[66] Yahfoufi N, Alsadi N, Jambi M, Matar C. The immunomodulatory and anti-inflammatory role of polyphenols. Nutrients. 2018;**10**(11). DOI: 10.3390/nu10111618

[67] Kwon O, Eck P, Chen S, Corpe CP, Lee JH, Kruhlak M, et al. Inhibition of the intestinal glucose transporter GLUT2 by flavonoids. FASEB Journal. 2007;**21**(2):366-377. DOI: 10.1096/fj.06-6620com

[68] Kinghorn AD, Su BN, Jang DS, Chang LC, Lee D, Gu JQ, et al. Natural inhibitors of carcinogenesis. Planta

Medica. 2004;**70**(8):691-705. DOI: 10.1055/s-2004-827198

[69] Shukla S, Kanwal R, Shankar E, Datt M, Chance MR, Fu P, et al. Apigenin blocks IKKα activation and suppresses prostate cancer progression. Oncotarget. 2015;**6**(31):31216-31232. DOI: 10.18632/oncotarget.5157

[70] Peng G, Du Y, Wei Y, Tang J, Peng AY, Rao L. A new synthesis of fully phosphorylated flavones as potent pancreatic cholesterol esterase inhibitors. Organic & Biomolecular Chemistry. 2011;**9**(7):2530-2534. DOI: 10.1039/C0OB00640H

[71] Gasiorowski K, Lamer-Zarawska E, Leszek J, Parvathaneni K, Yendluri BB, Błach-Olszewska Z, et al. Flavones from root of Scutellaria baicalensis Georgi: Drugs of the future in neurodegeneration? CNS & Neurological Disorders Drug Targets. 2011;**10**(2):184-191. DOI: 10.2174/187152711794480384

[72] Ganjare AB, Nirmal SA, Patil AN. Use of apigenin from Cordia dichotoma in the treatment of colitis. Fitoterapia. 2011;**82**(7):1052-1056. DOI: 10.1016/j.fitote.2011.06.008

[73] Wu X, Schauss AG. Mitigation of inflammation with foods. Journal of Agricultural and Food Chemistry. 2012;**60**(27):6703-6717. DOI: 10.1021/jf3007008

[74] Laev SS, Salakhutdinov NF. Anti-arthritic agents: Progress and potential. Bioorganic & Medicinal Chemistry. 2015;**23**(13):3059-3080. DOI: 10.1016/j.bmc.2015.05.010

[75] Nabavi SF, Braidy N, Gortzi O, Sobarzo-Sanchez E, Daglia M, Skalicka-Woźniak K, et al. Luteolin as an anti-inflammatory and neuroprotective agent: A brief review. Brain Research Bulletin. 2015;**119**(Pt A):1-11. DOI: 10.1016/j.brainresbull.2015.09.002

[76] Xiao J, Capanoglu E, Jassbi AR, Miron A. Advance on the flavonoid C-glycosides and health benefits. Critical Reviews in Food Science and Nutrition. 2016;**56**(Suppl. 1):S29-S45. DOI: 10.1080/10408398.2015.1067595

[77] Cardenas H, Arango D, Nicholas C, Duarte S, Nuovo GJ, He W, et al. Dietary apigenin exerts immune-regulatory activity in vivo by reducing NF-κB activity, halting leukocyte infiltration and restoring normal metabolic function. International Journal of Molecular Sciences. 2016;**17**(3):323. DOI: 10.3390/ijms17030323

[78] Shiota C, Abe T, Kawai N, Ohno A, Teshima-Kondo S, Mori H, et al. Flavones inhibit LPS-induced Atrogin-1/MAFbx expression in mouse C2C12 skeletal myotubes. Journal of Nutritional Science and Vitaminology (Tokyo). 2015;**61**(2):188-194. DOI: 10.3177/jnsv.61.188

[79] Park JB, Levine M. Intracellular accumulation of ascorbic acid is inhibited by flavonoids via blocking of dehydroascorbic acid and ascorbic acid uptakes in HL-60, U937 and Jurkat cells. The Journal of Nutrition. 2000;**130**(5):1297-1302. DOI: 10.1093/jn/130.5.1297

[80] Nakai M, Fukui Y, Asami S, Toyoda-Ono Y, Iwashita T, Shibata H, et al. Inhibitory effects of oolong tea polyphenols on pancreatic lipase in vitro. Journal of Agricultural and Food Chemistry. 2005;**53**(11):4593-4598. DOI: 10.1021/jf047814+

[81] Hsu CL, Yen GC. Induction of cell apoptosis in 3T3-L1 pre-adipocytes by flavonoids is associated with their antioxidant activity. Molecular Nutrition & Food Research. 2006;**50**(11):1072-1079. DOI: 10.1002/mnfr.200600040

[82] Liu IM, Tzeng TF, Liou SS, Chang CJ. Beneficial effect of traditional Chinese medicinal formula

Danggui-Shaoyao-san on advanced glycation end-product-mediated renal injury in streptozotocin-diabetic rats. Evidence-Based Complementary and Alternative Medicine. 2012;Article ID 140103. DOI: 10.1155/2012/140103

[83] Singab AN, Ayoub NA, Ali EN, Mostafa NM. Antioxidant and hepatoprotective activities of Egyptian moraceous plants against carbon tetrachloride-induced oxidative stress and liver damage in rats. Pharmaceutical Biology. 2010;**48**(11):1255-1264. DOI: 10.3109/13880201003730659

[84] Yang L, Takai H, Utsunomiya T, Li X, Li Z, Wang Z, et al. Kaempferol stimulates bone sialoprotein gene transcription and new bone formation. Journal of Cellular Biochemistry. 2010; **110**(6):1342-1355. DOI: 10.1002/jcb.22649

[85] Mahat MY, Kulkarni NM, Vishwakarma SL, Khan FR, Thippeswamy BS, Hebballi V, et al. Modulation of the cyclooxygenase pathway via inhibition of nitric oxide production contributes to the anti-inflammatory activity of kaempferol. European Journal of Pharmacology. 2010;**642**(1–3):169-176. DOI: 10.1016/j.ejphar.2010.05.062

[86] Boots AW, Drent M, De Boer VC, Bast A, Haenen GR. Quercetin reduces markers of oxidative stress and inflammation in sarcoidosis. Clinical Nutrition (Edinburgh, Scotland). 2011; **30**(4):506-512. DOI: 10.1016/j.clnu.2011.01.010

[87] Kobori M, Takahashi Y, Sakurai M, Akimoto Y, Tsushida T, Oike H, et al. Quercetin suppresses immune cell accumulation and improves mitochondrial gene expression in adipose tissue of diet-induced obese mice. Molecular Nutrition & Food Research. 2016;**60**(2):300-312. DOI: 10.1002/mnfr.201500595

[88] Dong YS, Wang JL, Feng DY, Qin HZ, Wen H, Yin ZM, et al. Protective effect of quercetin against oxidative stress and brain edema in an experimental rat model of subarachnoid hemorrhage. International Journal of Medical Sciences. 2014;**11**(3):282-290. DOI: 10.7150/ijms.7634

[89] Yang D, Liu X, Liu M, Chi H, Liu J, Han H. Protective effects of quercetin and taraxasterol against H_2O_2-induced human umbilical vein endothelial cell injury in vitro. Experimental and Therapeutic Medicine. 2015;**10**(4): 1253-1260. DOI: 10.3892/etm.2015.2713

[90] Miranda CL, Stevens JF, Ivanov V, McCall M, Frei B, Deinzer ML, et al. Antioxidant and prooxidant actions of prenylated and nonprenylated chalcones and flavanones in vitro. Journal of Agricultural and Food Chemistry. 2000; **48**(9):3876-3884. DOI: 10.1021/jf0002995

[91] Kumar A, Katiyar SB, Agarwal A, Chauhan PM. Perspective in antimalarial chemotherapy. Current Medicinal Chemistry. 2003;**10**(13): 1137-1150. DOI: 10.2174/0929867033457494

[92] Kontogiorgis C, Mantzanidou M, Hadjipavlou-Litina D. Chalcones and their potential role in inflammation. Mini Reviews in Medicinal Chemistry. 2008;**8**(12):1224-1242. DOI: 10.2174/138955708786141034

[93] Mojzis J, Varinska L, Mojzisova G, Kostova I, Mirossay L. Antiangiogenic effects of flavonoids and chalcones. Pharmalogical Research. 2008;**57**(4): 259-265. DOI: 10.1016/j.phrs.2008.02.005

[94] Qin L, Jin L, Lu L, Lu X, Zhang C, Zhang F, et al. Naringenin reduces lung metastasis in a breast cancer resection model. Protein & Cell. 2011;**2**(6): 507-516. DOI: 10.1007/s13238-011-1056-8

[95] Prabu SM, Shagirtha K, Renugadevi J. Naringenin in combination with vitamins C and E potentially protects oxidative stress-mediated hepatic injury in cadmium-intoxicated rats. Journal of Nutrional Science and Vitaminology (Tokyo). 2011;**57**(2):177-185. DOI: 10.3177/jnsv.57.177

[96] Céliz G, Daz M, Audisio MC. Antibacterial activity of naringin derivatives against pathogenic strains. Journal of Applied Microbiology. 2011; **111**(3):731-738. DOI: 10.1111/j.1365-2672.2011.05070.x

[97] Oršolić N, Gajski G, Garaj-Vrhovac V, Dikić D, Prskalo ZŠ, Sirovina D. DNA-protective effects of quercetin or naringenin in alloxan-induced diabetic mice. European Journal of Pharmacology. 2011;**656**(1–3):110-118. DOI: 10.1016/j.ejphar.2011.01.021

[98] Chao CL, Weng CS, Chang NC, Lin JS, Kao ST, Ho FM. Naringenin more effectively inhibits inducible nitric oxide synthase and cyclooxygenase-2 expression in macrophages than in microglia. Nutrition Research (New York, N.Y.). 2010;**30**(12):858-864. DOI: 10.1016/j.nutres.2010.10.011

[99] Bhardwaj P, Khanna D. Green tea catechins: Defensive role in cardiovascular disorders. Chinese Journal of Natural Medicines. 2013; **11**(4):345-353. DOI: 10.1016/S1875-5364(13)60051-5

[100] Norata GD, Marchesi P, Passamonti S, Pirillo A, Violi F, Catapano AL. Anti-inflammatory and anti-atherogenic effects of catechin, caffeic acid and trans-resveratrol in apolipoprotein E deficient mice. Atherosclerosis. 2007;**191**(2):265-271. DOI: 10.1016/j.atherosclerosis.2006.05.047

[101] Maruyama T, Tomofuji T, Endo Y, Irie K, Azuma T, Ekuni D, et al. Supplementation of green tea catechins in dentifrices suppresses gingival oxidative stress and periodontal inflammation. Archives of Oral Biology. 2011;**56**(1):48-53. DOI: 10.1016/j.archoralbio.2010.08.015

[102] Saito Y, Shimada M, Utsunomiya T, Imura S, Morine Y, Ikemoto T, et al. Green tea catechins improve liver dysfunction following massive hepatectomy through anti-oxidative and anti-inflammatory activities in rats. Gastroenterology. 2011;**140**(5):S-928. DOI: 10.1016/S0016-5085(11)63850-X

[103] Howes MJ, Simmonds MS. The role of phytochemicals as micronutrients in health and disease. Current Opinion in Clinical Nutrition and Metabolic Care. 2014;**17**(6):558-566. DOI: 10.1097/MCO.0000000000000115

[104] Weseler AR, Ruijters EJ, Drittij-Reijnders MJ, Reesink KD, Haenen GR, Bast A. Pleiotropic benefit of monomeric and oligomeric flavanols on vascular health—A randomized controlled clinical pilot study. PLoS One. 2011;**6**(12):e28460. DOI: 10.1371/journal.pone.0028460

[105] Suzuki-Sugihara N, Kishimoto Y, Saita E, Taguchi C, Kobayashi M, Ichitani M, et al. Green tea catechins prevent low-density lipoprotein oxidation via their accumulation in low-density lipoprotein particles in humans. Nutrition Research (New York, N.Y.). 2016;**36**(1):16-23. DOI: 10.1016/j.nutres.2015.10.012

[106] Esser D, Mars M, Oosterink E, Stalmach A, Müller M, Afman LA. Dark chocolate consumption improves leukocyte adhesion factors and vascular function in overweight men. FASEB Journal. 2014;**28**(3):1464-1473. DOI: 10.1096/fj.13-239384

Chapter 3

Rooibos (*Aspalathus linearis*) and Honeybush (*Cyclopia* spp.): From Bush Teas to Potential Therapy for Cardiovascular Disease

Shantal Windvogel

Abstract

Cardiovascular disease (CVD) is a leading cause of worldwide deaths. A number of risk factors for cardiovascular disease as well as type 2 diabetes and stroke present as the metabolic syndrome. Metabolic risk factors include hypertension, abdominal obesity, dyslipidaemia and increased blood glucose levels and may also include risk factors such as vascular dysfunction, insulin resistance, low high density lipoprotein (HDL) cholesterol levels and inflammation. Rooibos (*Aspalathus linearis*) and honeybush (*Cyclopia* spp.) are indigenous South African plants whose reported health benefits include anti-tumour, anti-inflammatory, anti-obesity, antioxidant, cardioprotective and anti-diabetic properties. The last two decades have seen worldwide interest and success for these plants, not only as health beverages but also as preservatives, flavourants and skincare products. This review will focus on the current literature supporting the function of these plants as nutraceuticals capable of potentially reducing the risk of cardiovascular disease.

Keywords: honeybush, rooibos, cardiovascular disease, diabetes, polyphenols

1. Introduction

Cardiovascular disease is the leading cause of deaths worldwide, killing 17.9 million people in 2016 [1]. While the number of cardiovascular disease related morbidity and mortality in the developed world has decreased or remained steady, the developing world has seen an increase. Limited resources, poverty, poor access to affordable healthcare, poor implementation of health policies, as well as poor education may be some of the reasons for the increase in cardiovascular diseases in low to middle income countries [2]. A number of risk factors for cardiovascular disease as well as type 2 diabetes and stroke present as metabolic syndrome. Metabolic risk factors include hypertension, abdominal obesity, dyslipidaemia, increased blood glucose levels, and may also include risk factors such as vascular dysfunction, insulin resistance, low levels of high density lipoprotein cholesterol (HDL-C) and inflammation. Natural products could play a significant role in drug discovery and development with examples including morphine, isolated from the opium poppy (*Papaver somniferum*) and artemisinin, from *Artemisia afra* [3–5]. Nutraceuticals are foods or supplements with health benefits [6]. To this end, a

IntechOpen

number of nutraceuticals including fruits, vegetables, tea and herbal infusions have shown health benefits. Approximately 80% of the emerging world relies on herbal supplements [7]. This may often be a more accessible form of health or self-care, due to a lack of access to modern medicine, an alternative to modern medicine or due to the high cost of treatment of modern medicine. This practice may involve the use of herbs or plants, including polyphenol rich rooibos (*Aspalathus linearis*) and honeybush (*Cyclopia* spp.), indigenous South African plant species with reported health benefits [8]. Many nutraceuticals contain polyphenols, the most abundant antioxidants in the diet which could help in the prevention of neurodegenerative diseases, diabetes, cancer, and cardiovascular disease [9]. Oxidative stress is a key process occurring in these diseases and is marked by imbalances between oxidants and the availability of antioxidants as well as perturbations in redox signalling mechanisms [10, 11]. Drugs used in the treatment of cardiovascular disease and obesity often have side effects, hence there is a need for better tolerated, safer and more natural treatment options [12]. A number of epidemiological studies and meta-analyses show some cardiovascular benefits with the intake of tea [13]. Furthermore, a review of some clinical studies show benefits of tea consumption in reducing cardiovascular risk factors, especially in overweight or obese subjects [14]. Rooibos and honeybush have a number of reported health properties, many of them targeting risk factors for the development of cardiovascular disease. The purpose of this paper was to review the role of rooibos and honeybush as potential nutraceuticals in the treatment of cardiovascular disease.

2. Risk factors for cardiovascular disease

A multitude of risk factors predispose to the onset of cardiovascular disease. This includes unmodifiable risk factors, such as increasing age, male gender, ethnicity, family history and genetics [15]. Modifiable risk factors include tobacco smoking, an unhealthy diet, a sedentary lifestyle, high alcohol intake, high blood pressure, being overweight or having central obesity, dyslipidaemia, impaired glucose tolerance or diabetes [15]. Diabetes not only quadrupled from 1980 to 2014 but approximately 57% of diabetic women and 67% of diabetic men are likely to present with cardiovascular disease by the age of 50 [1, 16]. Metabolic syndrome is largely preventable and includes a number of clinical findings which when occurring together, increase the risk of diabetes and cardiovascular disease. These include central obesity with any of the following risk factors including increased triglycerides, fasting plasma glucose, blood pressure and reduced HDL cholesterol levels [17]. Metabolic syndrome is also accompanied by changes in neuroendocrine and autonomic function [18]. It is known that early life stressors can predispose to disease outcome in later life, including cardiovascular disease [19]. Chronic stress influences cardiovascular outcome and anxiety and depression are also risk factors for cardiovascular disease [20–22]. This leads to changes in glucocorticoids and mineralocorticoids via modulation of the hypothalamic pituitary axis (HPA) [18]. Xenobiotics, including drugs and herbal infusions are metabolised by drug metabolising enzymes such as the phase I, cytochrome P450 system, which also influences the formation of steroid hormones [23]. CYP21A2 are precursors to both mineralocorticoids such as aldosterone and glucocorticoids such as cortisol and cortisone. Interestingly, rooibos flavonoids aspalathin and nothofagin inhibits CYP21A2 but not CYP11B1, which is responsible for converting 11-deoxycortisol to cortisol [24]. Substrate conversion of CYP117A1 and CYP21A2 was also inhibited by rooibos but other flavonoids such as rutin, orientin and vitexin were unable to inhibit CYP21A2. Rutin, under forskolin-induced stress, was the best inhibitor of steroid production followed by nothofagin and vitexin and lastly by aspalathin and nothofagin [24]. The observed effects for steroid inhibition were attributed to structural differences in these rooibos flavonoids. Rooibos also decreased rat glucocorticoids by decreasing the corticosterone, deoxycorticosterone as well as the corticosterone: testosterone ratio [25]. In the accompanying human study the cortisol: corticosterone ratio was reduced by rooibos, which also inhibited 11 β-hydroxysteroid dehydrogenase type 1 (11βHSD1). This enzyme catalyses the conversion of cortisone to cortisol and is associated with risk factors for cardiovascular disease [26]. In a stress model using steroid producing H295R cells, rooibos and rutin were able to reduce cortisol levels. The inhibition of mineralocorticoid and glucocorticoid steroids by rooibos and the dihydrochalcones aspalathin and nothofagin were also demonstrated in H295R cells [27]. This suggests that rooibos may offer a possible therapeutic role in the management of cardiovascular complications relating from stress, by altering the biosynthesis of steroid hormones via the HPA axis. Inhibition of 11βHSD1 has been suggested as a potential target mechanism for drugs to modulate the metabolic syndrome; therefore, this could be further explored in the context of rooibos or even honeybush. The cardiovascular complications of metabolic syndrome include coronary artery disease, peripheral vascular disease, hypertension as well as heart failure [28]. Primary management of metabolic syndrome is structured around lifestyle and dietary changes, including regular physical activity and a modest 5–10% initial reduction in caloric intake, failing which the use of pharmaceutical drugs may also be prescribed [17]. Approximately 1.9 billion people were overweight in 2016, of which 650 million people were obese [29]. A sedentary lifestyle, excess calories, a high fat diet and genetics contribute to the development of obesity, which is characterised by a body mass index greater (BMI) >30 kg/m^2 [30]. Urbanisation and a reduction in physical activity, along with an energy-rich Western diet, have contributed to the increase in cardiovascular disease in developing countries [31]. As a largely preventable disorder, obesity increases the risk of type 2 diabetes mellitus, cancer, cardiovascular disease, infertility, respiratory illnesses and a number of other health issues. In fact, obesity is not only a major independent risk factor but also an independent predictor for cardiovascular disease [32]. Strategies to avoid unnecessary deaths could include therapeutics that are safe, easily accessible and cost-effective. Rooibos and honeybush are relatively safe but long term clinical studies considering their safety are lacking [33, 34]. Two clinical case reports recommended caution in rooibos consumption but patients had consumed infusions containing rooibos and other herbs, thus the effects of these preparations and potential interactions between them need to be considered [35, 36]. Rooibos and honeybush are caffeine free and have low tannin levels making them ideal beverages for health conscious people, pregnant women and young children [37–39]. These plants may therefore be viable options in the future, provided sufficient evidence is generated to support their use as nutraceuticals capable of reducing cardiovascular risk.

3. Origin, distribution and markets for rooibos and honeybush

Rooibos [*Aspalathus linearis* L. (Burm.f.) R. Dahlgren (Leguminosae)] is a member of the fynbos biome, which contains needle-like leguminous plants. It occurs in the Cederberg area of the Western Cape, and in South Africa it is one of the most widely consumed herbal teas or tisanes. Its marketing potential was realized by Benjamin Gunzberg in 1904 and since then its popularity has steadily risen worldwide [40]. The top five export markets for rooibos are Germany, Japan, the Netherlands, the United Kingdom and the United States of America [41]. Rooibos is used as herbal infusion, health beverage, an ingredient in skin care products and cosmetics as well

Figure 1.
Honeybush plant in flower. Image courtesy of the SAHTA.

as a flavourant and colouring agent in a number of food applications. Honeybush (*Cyclopia* spp.), another member of the fynbos biome, is a bushy shrub found between the Piketberg area in the West, and Port Elizabeth in the East of South Africa (**Figure 1**). The year 1996 welcomed the first commercial harvests for honeybush, followed by the establishment of the South African Honeybush Tea Association (SAHTA) to manage the farming and sustainability practices as well as commercial interests of honeybush. After harvesting of rooibos or honeybush crops, leaves and stems are cut into small pieces, moistened and are fermented, either on open heaps or alternatively for honeybush also using an oven or fermentation tank. This is followed by drying of the fermented rooibos or honeybush. Fermentation of these plants is however associated with a change in phenolic composition as well as colour compared, to the green or unfermented plants which undergoes considerably less oxidation [42, 43]. The main contributors to the commercial market out of the 24 species of *Cyclopia* are *C. intermedia*, *C. genistoides* and *C. subternata*. *C. intermedia* has the largest market share; however, it is harvested from the wild, making the future sustainability and profitability of the crop problematic [44]. Honeybush is a budding commercial interest, used mainly as a tisane with great potential for development and is exported to countries such as the Netherlands, Germany and Japan [45].

4. Polyphenols that may be responsible for beneficial effects

Polyphenols are secondary plant metabolites commonly occurring in the diet in tea, coffee, wine, fruit, vegetables and cereals. The four main types of polyphenols, namely stilbenes, phenolic acids, flavonoids and lignins, can be classified according to the number of polyphenol rings and various chemical groups associated with the rings [46]. Flavonoids share a C3-C6-C3 backbone and the classification system includes groups such as the flavonols, flavones, isoflavones, flavanones, antho-cyanidins, and flavanols [46]. A number of studies have reported a reduction in the risk of cardiovascular disease with the intake of polyphenols [47]. The most prevalent polyphenols in rooibos include aspalathin, nothofagin, orientin, iso-orientin, vitexin and isovitexin, isoquercitrin and rutin [48, 49]. Aspalathin and aspalalinin are two unique dihydrochalcones in rooibos, with the former having been widely researched to date for its antioxidant and other health promoting properties [50, 51]. The flavonoid precursor in rooibos, Z-2-(β-D-glucopyranosyloxy)-3-phenylpropenoic acid (PPAG), has also received considerable attention for its anti-diabetic properties [52]. In *Cyclopia* species, the xanthones mangiferin, isomangiferin and the flavanone hesperidin are predominant [43, 53]. Mangiferin is not unique to honeybush, and also occurs in mangoes (*Mangifera indica*) and plants such as *Pyrrosia sheareri* and *Anemarrhena asphodeloides* [54–56]. Bioavailability refers to the amount of the substance that is ingested that is available for metabolism [46]. It involves a number of processes, including intestinal absorption, plasma kinetics, metabolism, binding to plasma albumin and excretion

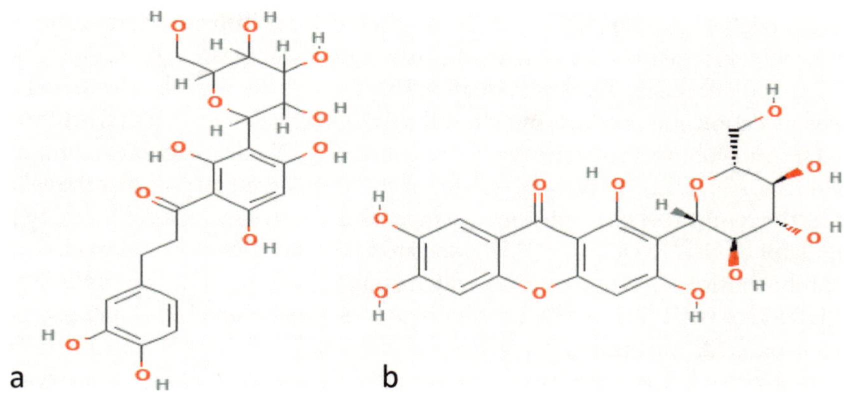

Figure 2.
Chemical structures of aspalathin (a) and mangiferin (b) showing phenol rings [62].

by the liver and kidneys [46]. Factors affecting bioavailability include isomeric form, processing methods, the type of compound and matrices surrounding the compound. The bioavailability of rooibos and honeybush is poor [57, 58]. The potential benefits attributed to their intake may therefore be hampered by poor bioavailability and also affect their maximal efficacy. Recently, the use of nanoencapsulation methods have been explored in order to increase the bioavailability and stability of aspalathin which could possibly promote the use of more effective nutraceuticals [59]. Since fermentation reduces polyphenol content, a number of studies have explored the benefits of unfermented, green rooibos or so called aspalathin rich extracts of rooibos in an attempt to elucidate the health promoting properties of the tisane [60, 61]. Others have looked at the effects of single isolate polyphenols from rooibos and honeybush such as aspalathin and mangiferin, to explore their functional benefits and identify a more targeted therapeutic option (**Figure 2**) [63, 64].

5. Antioxidant effects

Oxidative stress occurs due to an imbalance in the production of reactive oxygen species (ROS) and the availability of ROS scavengers. This may occur due to an excess of unstable reactive species including free radicals, such as reactive oxygen, reactive chlorine, reactive nitrogen and non-radical species, all of which may interact with cells, causing cellular damage [65]. Oxidative stress contributes to the pathobiology of cancer, cardiovascular disease, ageing, diabetes and atherosclerosis; therefore, it is plausible that dietary sources of antioxidants may be useful in reducing oxidative stress. Aerobic organisms are therefore reliant on innate antioxidant defences, e.g., superoxide dismutase (SOD), catalase (CAT), glutathione peroxidase (GPX), reduced glutathione (GSH), uric acid, albumin and peroxiredoxins, to deal with large quantities of ROS in an attempt to reduce oxidative stress. Mitochondria are key generators of ROS in aerobic organisms, producing them as they generate energy; however, ROS are also important in cell signalling and released by macrophages to promote immunological attacks [66]. Xenobiotics, including rooibos and honeybush, are metabolised by the cytochrome P450 family and in the process, radicals are also produced and further detoxification processes facilitate their removal [67]. In cardiovascular disease, oxidative stress may manifest as dysfunction in the vascular endothelium or cardiac myocytes, and may arise due to increased intracellular Ca^{2+} as a result of ROS [68]. Oxidative stress associated with hyperglycaemia may occur from the glycation of proteins and the formation of advanced glycation end products, the auto-oxidation of glucose and the polyol pathway.

In rodent models of streptozotocin (STZ)-induced diabetes, rooibos exerted antioxidant effects through increases in the activity of superoxide dismutase, catalase, glutathione peroxidase and decreasing lipid peroxidation [64, 69, 70]. Rooibos also decreased advanced glycation end products but clinical markers of diabetes such as fructosamine, glycated haemoglobin and glucose were unaffected [70]. Nuclear factor erythroid 2-related factor 2 (Nrf2), a critical regulator of the antioxidant response of the cell, facilitates the removal of oxidants through increased antioxidant enzyme activity [67]. The regulator of Nrf2, Kelch-like ECH-associated protein 1 (Keap 1), has oxidative and electrophilic sensitive cysteine residues which upon activation allows for dissociation and activation of Nrf2 [71]. In H9c2 cardiomyocytes, aspalathin (1 μM) increased the expression of Nrf2, and antioxidant genes and enzymes, including SOD, CAT, GPX and peroxiredoxins [64]. The expression of cytoprotective genes including heme oxygenase 1 (HO-1), NAD(P)H dehydrogenase (quinone 1), uncoupling protein 2 and apoptotic genes such as B-cell lymphoma 2 (Bcl-2) were also increased after aspalathin treatment. Uncoupling protein (UCP)3 and caspase 8, were however decreased. This suggests cellular survival due to reduced caspase 8, an important trigger for cell death [72]. UCP3 and UCP2 act in concert in the mitochondrial antioxidant response with UCP2 appearing to play a greater cytoprotective role in cardiomyocytes [73]. In the diabetic (db/db) mouse model, aspalathin increased Nrf2 expression as well as its downstream gene targets [74]. High dose aspalathin (130 mg/kg) also ameliorated the effects of hyperglycaemia in the heart by reducing the expected left ventricular enlargement. It could not however reduce fasting plasma glucose levels compared to metformin in these mice. The study confirmed the role of Nrf2 in the antioxidant response of the cell, strengthened the case for the potential use of nutraceuticals such as aspalathin in the treatment of diabetes, and also provides evidence for the differential effects observed with isolates compared to whole extracts of plants. Oxidative stress and inflammation co-exists in a number of diseases. The relationship between the two appears complex and is suggested to be a possible reason why antioxidant supplements in clinical trials have been unsuccessful [75]. In humans, no clear evidence of the antioxidant effects of flavonoids or pro-oxidant effects exist [76]. Supplementation of six cups of rooibos per day over a 6 week period in adults with increased cardiovascular risk increased plasma total polyphenols, reduced glutathione (GSH), and increased the GSH: oxidized glutathione (GSSG) ratio compared to the control period [77]. Furthermore, TBARS and conjugated dienes were reduced, indicating a reduction in oxidative stress after rooibos consumption. Ferric reducing antioxidant power (FRAP), oxygen radical absorbance capacity (ORAC) and 2,2′-azinobis-(3-ethylbenzothiazoline-6-sulfonate (ABTS) were unaffected. It is however known that assays for the measurement of antioxidant status can be difficult to compare and may also be non-specific and complicated by the instability of the species they are measuring [78]. Flavanoids and other polyphenols in honeybush are responsible for its antioxidant effects [79]. Fermentation of honeybush and rooibos however reduces antioxidant activity [80]. Mangiferin was the most effective scavenger of $ABTS^{\cdot+}$ and in terms of its ability to reduce ferric ion, than the flavanone eriocitrin or the flavone luteolin [43]. This could be due to the hydrophilic nature of mangiferin, which is a glucoside. In an *in vitro* study using skin cells, aqueous extracts of *Cyclopia subternata* sp. exhibited the highest $ABTS^{\cdot+}$ scavenging ability compared to other unfermented species of rooibos and *Camellia sinensis* [53]. In addition, rooibos, honeybush and Chinese green tea demonstrated ferric reducing antioxidant power (FRAP) and oxygen radical absorbance capacity (ORAC) abilities. Rooibos and honeybush also had better ORAC values than green tea, while rooibos aqueous extracts had the highest FRAP. In another model using hairless SKH-1 mice, unfermented honeybush and mangiferin had the highest FRAP compared to the fermented honeybush and hesperidin, and also the highest total antioxidant capacity [63]. Fermented and unfermented honeybush as well as mangiferin and hesperidin also protected against ultraviolet (UV) B-induced lipid peroxidation. Fermented and unfermented extracts of honeybush as well as hesperidin increased SOD and CAT activity, while mangiferin increased SOD but not catalase activity [63]. When Chinese green tea, rooibos and honeybush were evaluated for their ability to reduce lipid peroxidation, unfermented extracts generally offered better protection against lipid peroxidation but green tea offered the highest level of protection compared to the other infusions [53]. Mangiferin activated phosphatidylinositol 3-kinase (PI3K) induced protein kinase b (PKB/Akt) and Nrf2 signalling pathways, thus decreasing oxidative stress in an *in vivo* model using human kidney cells exposed to *tert*-butyl hydroperoxide [81]. The activities of SOD, CAT, GPX and the non-enzymatic intracellular antioxidant GSH were also enhanced and lipid peroxidation was decreased. When compared to vitamin C, ROS scavenging ability as measured by the 2,2-diphenyl-1-picrylhydrazyl (DPPH) radical scavenging method and FRAP was decreased at lower concentrations compared to vitamin C, which was used as the positive control. The expression of Nrf2, HO-1, SOD, Akt and GSK-β and the mechanistic target of rapamycin (mTOR) and cyclin D were also increased by exposure to mangiferin. In an STZ-induced diabetic Wistar rat model, mangiferin improved antioxidant status and reduced apoptosis and inflammation [82]. This could be due to modulation of the AGE-RAGE/MAPK signalling pathways. Advanced glycation end products (AGEs), which are increased in diabetes, lead to upregulation and activation of the receptor for advanced glycation end products (RAGE). This interaction causes inflammation and oxidative stress, through increases in the production of ROS, leading to lipid peroxidation. AGEs can also inhibit peroxisome proliferator activated receptor (PPAR)γ, a regulator in inflammation as well as the metabolism of lipids and glucose [83]. In the study mangiferin also displayed potential as a therapeutic in preventing AGE mediated lipogenesis. Antioxidants such as β-carotene, α-tocopherol and ascorbate were unsuccessful in reducing incidences of cardiovascular [84] and other diseases such as cancer [85], in fact appearing to increase risk in some cases, e.g., with vitamin C, x-tocopherol and beta-carotene [86–88]. Antioxidant supplementation to alleviate oxidative stress could however be affected by other factors such as the dose of the antioxidant, timing of the intervention, interactions with other antioxidants, whether it as administered as an isolate or a whole extract, the type of extract, the model as well as the methods that are used to detect oxidative stress. Rooibos and honeybush exert antioxidant effects by scavenging free radicals, chelating metal ions, or upregulating indigenous antioxidant enzymes. The aromatic ring structures of polyphenols also contain free hydroxyl groups which contribute to their antioxidant ability. The structural differences between rooibos polyphenols may thus also explain differences in antioxidant activity of these compounds [89]. The antioxidant activity of rooibos and honeybush polyphenols may also be explained by their ability to increase the activity of antioxidant enzymes, through upregulation of Nrf2, PI3K and other signalling pathways involved in cell survival. It may also be explained by their ability to act as antioxidants themselves, exerting differential physiological effects. The antioxidant effects of rooibos and honeybush and their polyphenols have been extensively studied in detail elsewhere and some of these effects are discussed here under the anti-inflammatory, anti-obesity, anti-diabetic and cardiovascular effects of these nutraceuticals (**Tables 1–4**).

6. Anti-inflammatory effects

Inflammation and oxidative stress form a common thread in the metabolic syndrome, type 2 diabetes and cardiovascular disease [132, 133]. The rooibos flavonoid orientin from rooibos reduced the number of mast cells in colon sections as well as

Plant species/compound	Model	Dosage	Mechanistic effects	Author and year
C. genistoides; *C. subternata*; *C. maculata*	Mesenteric lymph node cells; murine splenocytes	Various up to 250 μg/ml	Modulated immune response; increased IFN-γ, IL-4, $CD_4^+CD_{25}^+FOXP_3$ TREG cells: Total CD_4^+ cell ratio in mesenteric lymph node cells; increased IL-10 and IL-17a in splenocytes	Murakami et al., 2018 [90]
ASP; Noth	C57BL/6 mouse sepsis model	1 mg/kg BW	Decreased plasma blood urea nitrogen (BUN), creatinine, urine protein, LDH; inhibition of NF-κβ activation; decreased plasma nitric oxide (NO), TNF-α, IL-6, myeloperoxidase (MPO) and sepsis associated lethality; decreased oxidative stress by increasing kidney superoxide dismutase (SOD), catalase (CAT), glutathione peroxidase (GPx), and decreasing lipid peroxidation	Yang et al., 2018 [91]
Orientin	1,2-dimethyl hydrazine stimulated colorectal cancer in Wistar rats	10 mg/kg BW	Decreased inflammatory mast cells, NFκ-B, TNFX, IL-6, iNOS and cyclooxygenase-2 (COX)-2	Thangaraj and Vaiyapuri, 2017 [92]
C. intermedia; *C. subternata*	UVB/keratinocytes (HaCaT)	Various (0.09–0.1 mg/ml), aqueous extracts; 0.18-0.71/3 μg/ml, methanol extracts	Increased inhibition of cell viability, proliferation induced by UVB (aqueous *C. intermedia*); increased apoptosis, decreased intracellular interleukin (IL) 1-α (0.09–0.1 mg/ml); *C. subternata* (0.09–0.1 mg/ml) increased intracellular IL1-α and decreased extracellular IL-1α; methanol extracts alleviated reduction of cell growth parameters induced by UVB	Magcwebeba et al., 2016 [93]
Honeybush (fermented and scale up fermented honeybush extracts	HaCaT human keratinocyte cells exposed to UVB irradiation	10–100 μg/ml; 200 μg/ml (cell viability)	Anti-inflammatory: Decreased IL-1β, IL-6, IL-8; decreased ERK, P38, metallomatrix proteases (MMPs) and C-Jun N-terminal kinase (JNK); increased SOD, CAT activities	Im et al., 2016 [94]
ASP; Noth	LPS-induced HUVECs; C57BL/6 mice	Various up to 30 μM *in vitro*; ASP 27.1 μg/mouse, Noth 26.2 μg/mouse *in vivo*	Inhibition of LPS-induced barrier disruption; decreased expression of cell adhesion molecules (CAMs); decreased adhesion/transendothelial migration of leukocytes (HUVECs); decreased *in vivo* LPS-induced migration of leukocytes; decreased hyperpermeability; differentially decreased TNF-α, interleukin (IL)-6, NF-κB or ERK 1/2 by LPS; decreased LPS-induced lethal endotoxemia; increased antioxidant activity, decreased ROS; ASP inhibits effects on anti-inflammatory responses > Noth	Lee and Bae, 2015 [95]
C. subternata (scolymoside; vicenin-2)	HUVECs; C57BL/6 mice	Various up to 20 μM *in vitro*; 23.8 μg/mouse	Decreased adhesion and migration to HUVECS by human neutrophils *in vitro*, *in vivo*; decreased LPS release of transforming growth factor β-induced protein (TGFBIp); decreased TGFBIp mediated hyperpermeability; decreased TNF-α, IL-6, NF-κB, extracellular regulated kinases ½	Lee et al., 2015 [96]

Plant species/compound	Model	Dosage	Mechanistic effects	Author and year
C. subternata flavonoids: scolymoside (SCL); vicenin-2 (VCN)	Human umbilical vein endothelial cells (HUVECs); C57BL/6 mice	VCN 11.9 μg/mouse; SCL 23.8 μg/mouse (±20 μmol/L)	Decreased vascular permeability, monocyte adhesion, CAM expression, NFκ-B expression and ROS induced by high glucose; increased expression of SOD, CAT	Ku and Bae, 2015 [97]
Unfermented rooibos (uf) + methanol extracts; ASP, Noth	Non-steroidogenic transfected COS-1 cells, H295R cells	4.3 mg/ml; ASP (10 μm); Noth (10 μm)	Decreased steroid production; decreased glucocorticoids during forskolin treatment; decreased aldosterone and cortisol precursors (variable effects between individual polyphenols and extract)	Schloms et al., 2012 [27]
Honeybush (fermented, green ethanol soluble extracts); mangiferin (Mangif); hesperidin (Hesp)	SKH-1 mice	30 mg/ml extract; 3 mg/ml (Hesp); 4 mg/ml (Mang): 100 μl applied to dorsal skin	Anti-inflammatory: Fermented and unfermented extract decreased oedema, epidermal hyperplasia cyclooxygenase-2 (COX-2), ornithine decarboxylase (ODC), GADD45 and OGG1/2 expression; fermented extract decreased lipid peroxidation by increasing superoxide dismutase, catalase; isolated compounds hesperidin and mangiferin less effective than whole extracts	Petrova et al., 2011 [63]
Rooibos	Whole blood cultures unstimulated or stimulated with endotoxins or phytohemagglutinin PHA	250–7.8 μg/ml	Increased IL-6, 10, IFNγ (unstimulated cells); increased IL-6, decreased IL-10 (stimulated cells)	Hendricks and Pool, 2010 [98]
Rooibos	LPS stimulated macrophages	0.5 μg/ml	Decreased IL-6, IL-10; increased COX2 > 25%	Mueller et al., 2010 [99]
Rooibos	Wistar rats (Dextran sodium sulphate (DSS) induced rat colitis model)	1.6 g/100 ml BW *ad libitum*	Increased SOD vs. DSS rats; decreased 8-hydroxy-2′-deoxyguanosine (8-OHdG) vs. controls	Baba et al., 2009 [100]
Rooibos	Murine splenocytes	1–100 μg/ml	Increased ovalbumin; increased sheep RBC antibody production; no effect on specific LPS stimulated antibody response; increased IL-2 in ova anti-CD_3 primed splenocytes (10–100 μg/ml); decreased IL-4 in ova primed splenocytes; increased ova-induced antibody production in cyclosporine A rats; increased IL-2 in splenocytes	Kunishiro et al., 2001 [101]

Extracts are considered fermented unless otherwise indicated.

Table 1.
Anti-inflammatory and immune modulatory effects of rooibos and honeybush.

Plant species/ compound	Model	Dosage	Mechanistic effects	Author and year
Unfermented green rooibos extract (GRE)	Human C3A liver cells; obese insulin-resistant rats	10 µg/ml (cells); up to 195 mg/kg BW (6 cups), rats	Increased glucose and lipid metabolism: Increased glucose metabolism in C3A cells; increased insulin sensitivity in OBIR rats: increased GLUT2 expression; increased PI3K/Akt, phosphorylated AMPK and stimulation of insulin receptor substrate (IRS) 1, 2 forkhead box protein 01 (FOXO1) and carnitine palmitoyl transferase 1 (CPT1)	Mazibuko-Mbeje et al., 2019 [102]
ASP	Ventricular cardiomyocytes isolated from healthy, aged control and obese insulin resistant rats	10 µM	Increased insulin mediated glucose uptake in cardiomyocytes from young and aged rats, but not in high-caloric diet animals. Insulin actions enhanced via a PI3K-dependent mechanism	Smit et al., 2018 [103]
ASP; GRE	Palmitate-induced insulin resistant adipocytes	GRE (10 µg/ ml); ASP (10 µM)	Increased GLUT4 expression (GRE); Both treatments: decreased lipid-mediated insulin resistance; decreased NFκβ, IRS1 (ser^{307}), phosphorylated AMPK (GRE); increased Akt phosphorylation; Only ASP increased peroxisome proliferator-activated receptor (PPAR) α, γ and CPT1	Mazibuko et al., 2015 [104]
Fermented rooibos	3T3-L1 adipocytes	10 µg/ml, 100 µg/ml	Decreased lipid accumulation; decreased adipogenesis; decreased PPARγ, α, sterol regulatory binding factor 1 (SREBF1), fatty acid synthase (FASN) expression impaired; leptin secretion decreased	Sanderson et al., 2014 [105]

Plant species/ compound	Model	Dosage	Mechanistic effects	Author and year
Cyclopia maculata (aqueous)	3T3-L1 adipocytes	60–100 μg/ml	Anti-obesogenic 60–100 μg/ml fermented extracts increased glycerol release; 80 μg/ml increased lipolysis maximally; increased expression of perilipin, hormone sensitive lipase (HSL); not cytotoxic (up to 100 μg/ml)	Pheiffer et al., 2013 [106]
Cyclopia maculata (fermented, fermented), *Cyclopia subternata* (unfermented)	3T3-L1 pre-adipocytes	Various up to 1600 μg/ml	Anti-obesogenic: decreased adipocyte differentiation; decreased intracellular triglycerides (> 100 μg/ml); decreased cellular ATP (*C. Maculata*) decreased PPARγ isoform 2; increased adiponectin (fermented *C. maculata*); increased leptin cytotoxic: 800 μg/ml (*C. maculata unfermented*), 1600 μg/ml (*C. maculata+ C. subternata, unfermented*)	Dhudhia et al., 2013 [107]
Rooibos (not explicitly stated in paper but considered to be the fermented extract)	LDLr−/− mice; 3T3-L1 adipocytes	10 g/L (mice); 600 μg/ml (adipocytes)	Increased lipolysis; decreased serum cholesterol; triglycerides, free fatty acids; increased food consumption in mice fed normal chow; decreased body weight; altered adipocyte size, number; inhibition of dietary-induced steatosis; increased liver AMPK; decreased triglyceride accumulation; anti-adipogenic in 3T3-L1 adipocytes	Beltrán-Debón et al., 2011 [108]

Table 2.
Anti-obesity effects of rooibos and honeybush.

Plant species/compound	Model	Dosage	Mechanistic effects	Author and year
ASP	H9c2 Cardiomyocytes; db/db mice	1 μM (cardiomyocytes); (13 mg/kg BW)/130 mg/kg BW(mice)	Diabetic/cardioprotective/antioxidant; reduced high glucose induced oxidative stress; Nrf2-mediated activation of downstream antioxidant genes	Dludla et al., 2017 [64]
ASP	Glucose-exposed H9c2 cardiomyocytes	1 μM	Cardioprotective, anti-diabetic, antioxidant effects: Enhanced metabolism of glucose, decreased phosphorylation of AMPK, decreased CPT1; increased GLUT4, acetyl–CoA carboxylase expression; increased glutathione, SOD; decreased ROS, increased ucp2, bcl-2: bax; Anti-apoptotic; decreased DNA nicks	Johnson et al., 2016 [109]
Fermented *Cyclopia intermedia* (methanol fermented and scaled up extracts	HaCaT human keratinocyte cells	50–100 μg/ml	Antioxidant: Increased SOD, CAT in UVB-exposed HaCaT keratinocytes	Im et al., 2016 [94]
Cyclopia subternata (aqueous extracts); mangiferin, isomangiferin	STZ Wistar rat model; C2C12 cells	30–600 mg/kg (*in vivo*); 1 nM–100 μM (isolated compounds	Increased glucose tolerance 30, 60, 120 min (600 mg/kg BW; mangiferin, isomangiferin increased glucose uptake in C2C12 cells	Schulze et al., 2016 [110]
ASP; Green rooibos extract (GRE)	3T3-L1 adipocytes	100 μM; 10 μg/ml 10 μg/ml	Both isolates decreased palmitate-induced insulin resistance; decreased NFκ-β, IRS1 (Ser^{307}), phosphorylated AMPK; increased Akt activation; GRE increased GLUT4 expression; ASP increased PPARα, γ, CPT1 expression	Mazibuko et al., 2015 [104]
Unfermented rooibos	L6 myotubules; RIN-5F Cells; obese diabetic KK-Ay mice	350 μg/ml (L6 myotubules); 50 μg/ml (RIN-5F cells); 0.3–0.6% (mice)	Increased glucose uptake; increased phosphorylated AMPK, Akt; increased GLUT4 translocation; decreased AGE-induced ROS increase; decreased fasting blood glucose	Kamakura et al., 2015 [111]
Rooibos; Aspalathin (ASP); Nothofagin (Noth)	HUVECs; C57BL/6 mice	5–50 μM	Cardioprotective/anti-diabetic: Decreased vascular permeability caused by high blood glucose, decreased monocyte adhesion; Antioxidant: Decreased reactive oxygen species; Anti-inflammatory: Decreased NFκ-β	Ku et al., 2014 [112]

Plant species/compound	Model	Dosage	Mechanistic effects	Author and year
Fermented rooibos	H9c2 cardiomyocytes isolated from male Wistar STZ-induced diabetic rats (40 mg/kg BW); *in vivo*, *ex-vivo* rat heart perfusions	1, 10 μg/ml	Anti-diabetic/cardioprotective; decreased ROS, apoptosis; increased glutathione, metabolic activity *in vitro*	Dludla et al., 2014 [113]
Phenylpyruvic acid-2-*O*-β-D-glucoside (PPAG)	Obese mice	10 mg/kg	Protection from diet-induced hyperglycaemia; increased beta cell mass; decreased apoptosis; *In vitro*, protection of β-cells from palmitate-induced apoptosis; increased Bcl-2 expression; increased β-cell mass; protective effect of PPAG via increased expression of Bcl-2 in β-cells	Mathijs et al., 2014 [114]
Mangiferin	Diabetic insulin resistant Wistar rats	20 mg/kg BW	No change in body weight; decreased serum glucose; increased serum insulin; decreased HOMA IR; increased β-cell function; decreased serum TNFα; improved serum lipids; increased adiponectin; no effect on antioxidant activity	Saleh et al., 2014 [115]
Cyclopia maculata (*unfermented*); Mangiferin	RIN-5F cells	0.001–1000 μg/ml (*C. maculata*); 0.01–1000 μg/ml (mangiferin)	Increased viability (0.001–1000 μg/ml) *Cyclopia*; no effect on viability (mangiferin)	Chellan et al., 2014 [116]
Cyclopia maculata	STZ-diabetic Wistar rats	30/300 mg/kg BW	Increased glucose tolerance; decreased fasting blood glucose; improved serum triglycerides; increased β-cell area; increased β-cell proliferation (300 mg/kg BW); decreased plasma nitrite; unaltered catalase, glutathione, liver lipid peroxidation and nitrotyrosine	
Z-2-(β-D-glucopyranosyloxy)-3-phenylpropenoic acid (PPAG)	Chang cells; Obese insulin-resistant rats	1–31.6 μM; 0.3–3 mg/kg BW (obese rats)	Increased glucose uptake; decreased fasting blood glucose; increased glucose tolerance; increased mRNA expression of liver GLUT1, 2, glucokinase, PPARγ and SOCS3	Muller et al., 2013 [52]

Plant species/compound	Model	Dosage	Mechanistic effects	Author and year
ASP	L6 myocytes; RIN-5F cells; type 2 diabetic ob/ob mice	25–100 μM 0.1% ASP (100 mg/kg BW); 10 mg/kg BW ASP (intraperitoneal glucose tolerance test)	Increased glucose uptake; increased AMPK phosphorylation; increased GLUT4 translocation in L6 myoblasts and myotubules; decreased age-induced ROS; decreased blood glucose; increased glucose tolerance; decreased expression of liver gluconeogenic and lipogenic gene expression in mice	Son et al., 2013 [117]
Unfermented green rooibos; fermented rooibos	C2C12 skeletal muscle cells	10 μg/ml	Increased glucose uptake; increased mitochondrial activity; increased ATP (unfermented >fermented); decreased PKCq activation; decreased palmitate-induced insulin resistance; increased insulin-dependent Akt activation, increased AMP Insulin-independent signalling pathways; increased GLUT4	Mazibuko et al., 2013 [118]
ASP; rutin	C2C12 myotubules	1,10,100 μM; 100 μm	Increased glucose uptake	Muller et al., 2012 [119]
Aspalathin: rutin	STZ diabetic rats	1:1 (1.4 mg/kg BW)	Decreased blood glucose (not obtained by individual compounds)	
Aspalathin rich green rooibos	STZ diabetic rats	25 mg/kg BW, 30 mg/kg BW	Decreased blood glucose; increased glucose tolerance	
Cyclopia intermedia	STZ diabetic induced Wistar rats	Acute: 5 mg/kg bw-50 mg/kg BW (STZ rats)	Decreased fasting blood glucose (50 mg/kg)	Muller et al., 2011 [120]
	Obese insulin resistant (OBIR) rats	Chronic: 538–2688 mg/ml (OBIR rats)	Decreased plasma cholesterol, fasting blood glucose; α:β cell mass (538–2150 mg/ml); decreased glucose tolerance (1075–2688 mg/ml)	
ASP from GRE	L6 myotubules	1–100 μM	Increased glucose uptake	Kawano et al., 2009 [121]
	RIN-5F cells	100 μm	Increased insulin secretion	
	db/db mice	0.1–0.2%	Decreased fasting blood glucose for 5 weeks; increased glucose tolerance 30, 60, 90, 120 min	
Fermented rooibos (aqueous + alkaline extracts)	STZ-induced diabetic Wistar rats	2.5% *ad libitum*	Decrease AGE's + MDA (plasma, lens, liver and kidney); decreased total cholesterol and creatinine	Uličná et al., 2006 [70]

Table 3.
Anti-diabetic effects of rooibos and honeybush.

Plant species/compound	Model	Dosage	Mechanistic effects	Author and year
Aspalathin	H9c2 cardiomyocytes; db/db mice	1 µm (cardiomyocytes) (13 mg/kg)/130 mg/kg	Diabetic/cardioprotective/antioxidant: reduced high glucose induced oxidative stress. Nrf2 activated downstream antioxidant genes; high dose aspalathin treatment was more successful than metformin or lower dose aspalathin at activating Nrf2 and antioxidant genes	Dludla et al., 2017 [64]
Aspalathin	H9c2 cardiomyocytes exposed to glucose	1 µM	Cardioprotective/anti-diabetic: Enhanced metabolism of glucose, decreased 172 phosphorylated AMPK, decreased carnitine palmitoyltransferase (CPT)1, increased GLUT4, acetyl-CoA carboxylase expression; antioxidant effects: increased glutathione and SOD, decreased ROS, increased UCP2, Bcl-2:Bax; anti-apoptotic; decreased DNA nicks	Johnson et al., 2016 [109]
Fermented rooibos	H9c2 cardiomyocytes isolated from male Wistar STZ diabetic rats (40 mg/kg BW)	1/10 µg/ml	Decreased ROS, apoptosis, increased glutathione, and metabolic activity *in vitro*	Dludla et al., 2014 [113]
Luteolin	STZ diabetic induced Wistar rats	10 µ/kg BW	Decreased LDH, incidences of arrhythmias; decreased infarct size; increased left ventricular ejection fraction; decreased myocardial apoptosis; increased FGFR2, LIF expression; increased PI3K/Akt, Bcl-2 associated death promoter (BAD), Bcl-2:Bax ratio; inhibited MPO expression, IL-6, IL-1a and TNFα production (anti-inflammatory effects)	Sun et al., 2012 [122]
Luteolin	Ischemia/reperfusion (I/R) Wistar rats and isolated cardiomyocytes	Various up to 40 µM, model dependant	Luteolin pre-treatment before I/R injury increased contractions in isolated rat heart and cardiomyocytes; decreased LDH, apoptosis; increased Bcl-2:Bax ratio; decreased infarct size; protection involves ERK 1/2-PP1a-PLB-SERCA2a mechanism	Wu et al., 2013 [123]
Rooibos	Human volunteers	400 ml (0.025 g/ml)	Single dose inhibited ACE after 30, 60 min; ACE II genotype inhibited after 60 min; no effect on nitric oxide	Persson et al., 2010 [124]
Rooibos	HUVEC	0.05 g/ml in phosphate buffered saline (PBS)	Increased nitric oxide after 1 day (1:400, 1:200); no effect on ACE inhibition after 10 min	Persson et al., 2006 [125]
Rooibos	Human serum with Enalaprilat as positive control	0.05 g/ml in PBS	Inhibition of ACE via a mixed inhibition method	Persson et al., 2012 [126]

Plant species/compound	Model	Dosage	Mechanistic effects	Author and year
Rooibos	Human volunteers at risk for cardiovascular disease	0.005 g/ml (6 cups/day)	Increased antioxidant status and decreased oxidative stress by increased GSH: GSSG ratio, decreased lipid peroxidation (TBARS, conjugated dienes); increased HDL cholesterol	Marnewick et al., 2011 [77]
Orientin	Ischemia/reperfused Wistar rat hearts; cardiomyocytes injured by hypoxia/ reoxygenation	0.5–2 mg/kg BW (rat hearts); 3–30 μmol/l (cardiomyocytes)	Decreased apoptosis in cardiomyocytes; increased Bcl-2, decreased bcl-2-like protein 4 (Bax), increased Bcl-2: Bax; decreased cytochrome-c, caspase expression in cardiomyocytes and myocardium	Fu et al., 2006 [127]
Orientin	Ischemia/reperfusion injured H9c2 cardiomyocytes	30 μM	Decreased apoptosis (decreased MPTP opening)	Lu et al., 2011 [128]
Rooibos (fermented and unfermented)	*Ex vivo* working heart from Wistar rats	2% w/v	Increased aortic output; decreased cleaved caspase 3, poly (ADP-ribose) polymerase (PARP), reduced apoptosis; increased GSH: GSSG ratio	Pantsi et al., 2011 [129]
Rooibos; chrysoeriol	Sprague Dawley rats (blood pressure effects); rabbit aorta, guinea pig atria	20% w/w, 10–100 mg/ kg BW	Glibenclamide sensitive relaxation of low K^+-induced contractions; weak inhibitory effect on atrial force and contraction rate; blood pressure lowering effects; chrysoeriol caused concentration-dependent, glibenclamide-sensitive relaxation of low K^+ induced contractions, EC50 (61 μg/ml), n = 2 in aorta	Khan and Gilani, 2006 [130]
Rooibos and honeybush flavonoids (various including luteolin, quercetin, rutin, genestein, hesperitin, etc.)	*In vitro* ACE-inhibition assay	500 μM, 100 μM	ACE inhibitory activity of some flavonoids found in rooibos and honeybush. Luteolin had the greatest ACE inhibitory effects compared to the other flavonoids with hesperitin and genestein exhibiting moderate but lower, yet marked effects	Guerrero et al., 2012 [131]

Table 4.
Cardiovascular effects of rooibos and its flavonoids.

NFK-β, TNF-α, IL-6, iNOS and COX2, reflecting its anti-inflammatory potential [92]. Aspalathin and nothofagin inhibited LPS-mediated expression of the lipopolysaccharide (LPS) receptor (TLR4) and LPS-mediated barrier disruption. This occurred by increasing barrier integrity and inhibiting the expression of cell adhesion molecules (CAMs) and reducing neutrophil adhesion and migration in human umbilical vein endothelial cells [95]. The barrier protective effects of aspalathin and nothofagin were confirmed in a mouse model, in which the dihydrochalcones reduced LPS-induced mortality. This suggests that aspalathin and nothofagin can be regarded as potential therapeutics against vascular inflammation. The protective effects of aspalathin and nothofagin in a systemic inflammatory response induced by caecal ligation included decreases in inflammatory markers, including TNFα-expression NF-κB, NO and IL-6 production, enhanced antioxidant activity and decreased lipid peroxidation. [91]. The anti-inflammatory effects of rooibos and honeybush may involve a reduction in inflammatory cytokines such as IL-6, TNF-α, COX2 [91, 92, 94]. It may also involve a reduction in neutrophil adhesion, and other mechanisms [96], **Table 1**.

7. Anti-obesity effects

Obesity and overweight are major risk factors for cardiovascular disease and type 2 diabetes. Insulin resistance, a key finding in the metabolic syndrome, is characterised by alterations in glucose uptake in insulin sensitive tissues such as the liver, skeletal muscle and adipose tissue [134]. Non-alcoholic fatty liver disease (NAFLD), characterised by fat accumulation in the liver, is an independent predictor of the metabolic syndrome which influences the progression of cardiovascular disease [135]. Since NAFLD is complex, utilising therapeutic strategies with multiple targets may be beneficial. The PI3K/Akt pathway is a complex insulin regulated pathway involved in glucose and lipid metabolism, as well as other cellular processes, including protein synthesis, cell signalling cell growth and apoptosis [136]. In type 2 diabetes and obesity, inhibition of the pathway interferes with beta cell function and insulin secretion, exacerbating insulin resistance. A green rooibos extract reduced lipid accumulation as well as lipolysis in C3A liver cells [102]. Rooibos also ameliorated palmitate-induced insulin resistance, by activating phosphorylated PKB/Akt and 5′ adenosine monophosphate-activated protein kinase (AMPK), as well as increasing glucose transporter (GLUT)2 expression. Furthermore, fatty acid oxidation was enhanced by increasing FOXO1, decreasing malonyl-CoA decarboxylase, increasing carnitine palmitoyl transferase 1 (CPT1) and increasing acetyl CoA carboxylase in insulin deficient cells. In the accompanying obese insulin resistant rat model, only a high dose (195 mg/kg) of green rooibos extract (GRE) could reduce insulin levels and the HOMA-IR index. No significant differences were detected in blood glucose, body weights or food intake. Insulin receptor and insulin receptor substrates 1, and 2 were however upregulated by GRE, suggesting that GRE may be beneficial in ameliorating obesity-induced insulin resistance. In 3T-L1 adipocytes, palmitate-induced insulin resistance was ameliorated by both aspalathin, as well as GRE [104]. This was accompanied by increases in Akt activation and decreases in nuclear factor (NF)-κB, IRS1 and AMP phosphorylation. GLUT4 expression was enhanced by the GRE but not aspalathin, suggesting that the whole extract rather than the isolated compound may be beneficial in providing a more multi-targeted therapeutic approach in improving palmitate-induced effects on glucose and lipids. A number of other rooibos polyphenols, such as the C-glycosidic flavones orientin, iso-orientin and luteolin, have anti-obesity effects, as reflected by their inhibition of pancreatic lipase, which is responsible for digesting and absorbing triglycerides [137]. Fermented rooibos also inhibited lipid accumulation in 3T3-L1 adipocytes [105]. Z-2-(β-D-glucopyranosyloxy)-3-phenylpropenoic acid (PPAG), increases glucose uptake, as demonstrated in Chang cells [52]. In their *in vivo* model of obese insulin resistant rats, basal fasting glucose decreased and glucose tolerance was improved by PPAG. Increases in mRNA expression of genes involved in glucose (GLUT1 and GLUT4) and insulin metabolism (IR) and others such as PPARα and SOCS3 in the liver were also seen. PPARα regulates the action of fatty acids in the liver, suggesting that insulin signalling, glucose and lipid metabolism may be altered by PPAG. SOCS3 associates with various proteins to inhibit cytokine signals, e.g., leptin, growth hormone, IL-6, leukaemia inhibitory factor (LIF) as well as insulin [138]. It may therefore mediate leptin resistance, which occurs in obesity but is also regulated by leptin, which may partially explain these findings. Administration of rooibos to LDL receptor deficient ($Ldlr^{-}/^{-}$) mice reduced liver steatosis and white adipose tissue and increased brown adipose tissue in high fat diet rats. Macrophage recruitment was decreased but no difference was seen in adipocyte size or number in the rooibos supplemented group. Furthermore, there appeared to be no liver toxicity, and free fatty acids, triglycerides and cholesterol were reduced by rooibos in the high fat diet group [108]. In humans, rooibos reduced oxidative stress, improved HDL-C, and triacylglycerol and low density lipoprotein cholesterol (LDL-C) in adults at risk of developing cardiovascular disease. [77]. These results suggest that rooibos and flavonoids in rooibos, such as aspalathin, may be potential adjuvants in the management of the metabolic syndrome. The anti-obesity effects of *Cyclopia intermedia*, *C. subternata* and *C. maculata* are limited but have been studied in 3T3-L1 adipocytes [106, 107, 139]. *C. maculata* and *C. subternata* decreased intracellular lipid and triglycerides and increased PPARγ, a regulator of glucose and lipid metabolism [107]. Their anti-obesity effects also include increased release of glycerol [106]. Hormone sensitive lipase (HSL) expression, which is regulated by SIRT1a, a rate limiter of lipolysis, was also increased by fermented *C. maculata*. Furthermore, perilipin expression, which appears to play a key role in the metabolism of lipids and lipolysis of adipose tissue was upregulated [140]. *C. intermedia* increased HSL, SIRT1, UCP3 as well as PPARγ expression in 3T3-L1 adipocytes [139]. The authors attributed increases in SIRT1 and PPARγ to be indicative of changes from white to brown adipose tissue, however the expression of UCP1, which characterises brown adipose tissue was not measured in the study [141]. While the exact function is unclear, UCP3 may protect against lipotoxicity of the mitochondria [142]. In the parallel obese db/db mouse model, no increases in body weight gain were seen with *C. intermedia* treatment, and neither food nor water consumption was affected [139]. Extracts of mangiferin and other polyphenols isolated from plants containing components similar to that found in honeybush have proven useful in understanding their mechanisms of action and how they could be used as nutraceuticals in the treatment of cardiovascular disease. In a double-blind randomised placebo controlled clinical trial, mangiferin (150 mg/day) from mangoes, was given to 97 overweight, hyperlipidaemic patients over a period of 12 weeks. Increased high density lipoprotein cholesterol (HDL-C), L-carnitine, as well as decreases in total cholesterol, low density lipoprotein cholesterol (LDL-C) and triglycerides were obtained in the patients [143]. No alterations in liver and kidney function markers were seen, suggesting that chronic treatment over 12 weeks was safe in human participants. Since the metabolic syndrome is so complex, providing a multi-targeted approach, such as through affecting the PI3K/Akt pathway may be somewhat beneficial. The anti-obesity effects are further described in **Table 2**. The limited available information on the anti-obesity effects of rooibos and honeybush however, suggests that more research is needed to fully understand and enable the translation of these effects from animals to man.

8. Anti-diabetic effects

Diabetic hyperglycaemia involves macro-and microvascular complications [144]. Free radicals are also produced, promoting oxidative stress [145]. Evidence suggests that rooibos and its polyphenols may reduce oxidative stress associated with the pathogenesis of diabetes [34, 69]. A meta-analysis and systematic review narrowed down to 12 peer-reviewed studies reported a reduction in blood glucose levels in diabetic rodent models that had received either rooibos or its polyphenols, while no clinical trials have investigated the effects of rooibos in the context of type 2 diabetes [146]. Rooibos decreases fasting blood glucose and improves glucose tolerance [117]. The anti-diabetic actions may be related to the expression of genes responsible for glucose uptake (GLUT1, GLUT2), insulin signalling (IR), as well as other genes such as PPRα in the liver in a model of obese insulin resistant rats treated with PPAG [52]. The involvement of insulin dependent GLUT4, in skeletal muscle has also been reported [111]. Rooibos may be exerting anti-diabetic effects by affecting the metabolism and uptake of glucose. Aspalathin influences key genes involved in the metabolism of lipids, insulin resistance, inflammation and apoptosis, possibly reversing metabolic abnormalities by involving PPARγ, Adipoq, IL-6/Jak2 pathway and Bcl-2 [74]. PPAG increased beta cell mass and delayed hyperglycaemia in STZ-diabetic mice [147]. This was accompanied by increases in anti-apoptotic B-cell lymphoma 2 (Bcl-2), but no antioxidant effect, and in human islet cells PPAG also decreased cell death. This suggests that PPAG increases beta cell mass by decreasing apoptosis and increasing Bcl-2 as reflected in both STZ diabetic and obese mouse models [114, 148]. In the absence of insulin, GRE also increased phosphorylation of both AMPK and Akt, in L6 skeletal myotubules. Activation of AMPK could be a therapeutic strategy in the treatment of obese and type 2 diabetics, as animal studies suggest a dysregulation of AMPK in these states. The mechanism for the amelioration of diabetic complications in rodents by rooibos and honeybush may also be due α-glucosidase inhibition; or SGLT2 inhibiting potential of rooibos and honeybush or their respective flavonoids [119, 149–151]. α-Glucosidase is an enzyme present on the intestinal brush border where it digests starch, increasing blood glucose [152]. Anti-diabetic drugs, therefore commonly utilize strategies such as α-glucosidase inhibition or SGLT2 inhibition to facilitate their actions [152, 153]. The dihydrochalcone phlorizin, an SGLT2 inhibitor, provides a basis to explore natural plant based agents for use in the management of diabetes [154]. Strategies to target SGLT2 rather than SGLT1 is also associated with reduced drug toxicity as found in studies of phlorizin and anti-diabetic drugs [153]. In an STZ-diabetic model, unfermented extracts of *Cyclopia maculata* improved glucose tolerance, fasting glucose, β-cell area, triglyceride levels as well as the insulin: glucagon ratio. Plasma nitrite was reduced but no alterations were seen in other markers of nitrotyrosine and lipid peroxidation or the serum antioxidant enzymes [155]. Mangiferin and naringenin from *Salacia oblonga* also reduced blood glucose, normalised AST and ALT levels, improved antioxidant status and decreased protein carbonyl levels, glycogen and TBARS in the liver. Beta cell damage in the pancreas was also ameliorated by the compounds. It was proposed that activation of PPARγ and GLUT4, which was also accompanied by insulin sensitisation, may be partially responsible for the anti-diabetic effects [156]. Since naringenin is also found in honeybush, it suggests that extracts from honeybush could possibly also produce similar effects. Mangiferin from *Anemarrhena asphodeloides Bunge* reduced blood glucose levels and ameliorated hyperinsulinemia, the anti-diabetic properties possibly due to a reduction in insulin resistance in these rats [157]. Furthermore, mangiferin and naringenin isolated from *Salacia oblonga* reduced blood glucose levels, and increased GLUT4 and PPARγ expression, suggesting increased insulin sensitisation and demonstrating the anti-diabetic properties of the polyphenols. In conclusion, the anti-diabetic effects of rooibos and honeybush (**Table 3**) are mediated by complex signalling pathways that affect glucose and lipid metabolism as well as oxidative stress.

9. Cardiovascular effects

Angiotensin converting enzyme (ACE) inhibitors reduce blood pressure by inhibiting the conversion of angiotensin to angiotensin II, the latter being a potent vasoconstrictor which increases blood pressure through a variety of mechanisms [158]. Angiotensin II can also interact with the angiotensin I receptor, enhancing ROS production, which contributes to endothelial dysfunction by inactivating vasodilatory nitric oxide [158]. Endothelial dysfunction is commonly associated with risk factors of cardiovascular disease, including obesity and type 2 diabetes [159]. It also precedes atherosclerosis, which involves a number of complex processes such as the release of pro-inflammatory cytokines and inflammation, oxidation of LDL as well as macrophage recruitment and platelet adhesion [158]. Cardiovascular protective effects of rooibos have also been established in a number of studies (**Table 4**). In an *in vitro* human umbilical vein endothelial cell (HUVEC) study, rooibos could not inhibit ACE but increases in nitric oxide were detected [125]. ACE-inhibition was also determined in 17 healthy volunteers receiving a single oral dose of rooibos [124]. Inhibition of ACE occurred up to 60 min after ingestion but no change in blood pressure or nitric oxide was detected. ACE-inhibitory effects have also been reported for rooibos and honeybush and some of their flavonoids [131]. This effect on ACE could be due to the presence of the C2═C3 double bond, the catechol group at the B-ring and the cetone group on C4 on the C-ring of flavonoids [131]. The best ACE-inhibitor in HUVEC's was luteolin, followed by quercetin, rutin, kaempferol, rhoifolin and apigenin K. The honeybush flavonoids genistein and hesperetin also displayed ACE-inhibitory effects but these were less potent than luteolin. Rooibos probably inhibits ACE using a mixed type of inhibition, which binds to the enzyme's active and allosteric sites [126]. Organic fractions of, as well as flavonoids in rooibos such as chrysoeriol and rutin inhibited spontaneous and low and/or high K^+-induced contractions in models utilising jejunum and bronchial smooth muscle [130, 160]. Anti-contractile effects were reportedly due to the opening of K^+ channels as well as a Ca^{2+} antagonistic action. These studies were not done in vascular smooth muscle, but it is possible that rooibos could exert antihypertensive effects via similar mechanisms in vascular smooth muscle. It also caused mild reduction of atrial force and rate of spontaneous contractions in the Guinea pig and decreased blood pressure in anaesthetised rats [130]. In comparison, isolated compounds of rooibos tested in jejunal, tracheal and aortic preparations showed variable effects. In adults with cardiovascular risk factors, six cups of rooibos was able to increase antioxidant status by increasing the GSH: GSSG ratio and decrease lipid peroxidation by decreasing TBARS and conjugated dienes [77]. Furthermore, increases in the healthy HDL cholesterol were also seen. In a recent study using a high fat diet to induce obesity in Wistar rats, treatment with an enriched GRE, containing large quantities of aspalathin, improved glucose tolerance, vascular function, as determined by aortic vascular contraction-relaxation studies and enhanced the antioxidant status. Modest improvements in blood pressure were also seen in a recent study in obese rats, suggesting cardiovascular benefits with the intake of the rooibos [161]. Fermented rooibos also improved vascular function in nicotine exposed rats [162]. Since tobacco smoking is a major risk factor for cardiovascular disease, this suggests a possible therapeutic role for rooibos in the reduction of cardiovascular complications associated with tobacco smoking, such as vascular dysfunction or oxidative stress. Cardioprotective effects of rooibos and its flavonoids typically appear to involve a reduction in lipid peroxidation and upregulation of the antioxidant enzymes through Nrf2 activation, and a reduction in apoptosis which may also be associated with increases in the Bcl-2: Bax ratio [77, 122, 129]. The anti-inflammatory effects of rooibos, also support its role as a cardioprotective agent (**Table 1**). No literature investigating the vascular modulating effects of honeybush are currently available and anti-inflammatory effects on HUVECs are discussed above. Decreases in VLDL, LDL-C and increases in HDL-C by naringenin and mangiferin from *Salacia* sp. however suggests the potential of similar isolates from honeybush to ameliorate endothelial dysfunction [156].

10. Conclusions and perspectives

Rooibos and honeybush have a number of health benefits. The evidence for their use as functional nutraceuticals is reflected in anti-diabetic, anti-inflammatory, antioxidant, anti-obesity, and vascular effects in animal and *in vitro* studies. The translation of these results into human medical care has however been limited. The inability to translate animal studies to that of humans may be due to the result of the controlled genetic background and environmental conditions of animals used as models to study disease [163]. Given the unstable nature of ROS and the limitations in assays used to detect oxidative stress, it is noteworthy to remember that intervention studies may only as good as the biomarkers used to measure the intervention. As we investigate the therapeutic role of rooibos or honeybush in the treatment of cardiovascular disease, it is evident that we are only but scratching the surface. These plants have complex chemical compositions and while isolating single components could be beneficial, the interplay and additive effects of other components in whole extracts could be more beneficial [111]. The mechanisms behind these benefits involve numerous signalling pathways that regulate their effects. Understanding these complexities may be what is needed to promote their transition from bush teas to nutraceutical therapy against cardiovascular disease but until then we may as well enjoy a cuppa.

Conflict of interest

There is no conflict of interest to disclose.

Thanks

Thanks go to Professor H. Kuivaniemi for proofreading this manuscript and to the SAHTA for allowing the use of the honeybush image.

Author details

Shantal Windvogel
Stellenbosch University, Cape Town, South Africa

*Address all correspondence to: shantalw@sun.ac.za

References

[1] World Health Organization. Noncommunicable diseases Factsheet No. 355. Geneva, 2018 [cited 2019 Apr 15]. Available from: http://www.who.int/mediacentre/factsheets/fs355/en/

[2] Agyemang C, van den Born B-J. Limited access to CVD medicines in low-income and middle-income countries: Poverty is at the heart of the matter. The Lancet Global Health. 2018;**6**(3):e234-e235

[3] Newman DJ, Cragg G. Natural products as sources of new drugs over the 30 years. Journal of Natural Products. 2012;**75**(3):311-335

[4] Krishnamurti C, Rao SSCC. The isolation of morphine by Serturner. Indian Journal of Anaesthesia. 2016;**60**(11):861-862

[5] Tu Y. Artemisinin—A gift from traditional Chinese medicine to the world (Nobel lecture). Angewandte Chemie, International Edition. 2016;**55**(35):10210-10226

[6] DeFelice SL. Foundation for innovation in medicine rationale and proposed guidelines for the nutraceutical research & education act. Journal of Nutraceuticals Functional & Medical Foods. 2002;**2**(1):43-52

[7] Ekor M. The growing use of herbal medicines: Issues relating to adverse reactions and challenges in monitoring safety. Frontiers in Pharmacology. 2014;**4**:1-10

[8] Mahomoodally MF. Traditional medicines in Africa: An appraisal of ten potent African medicinal plants. Evidence-based Complementary and Alternative Medicine. 2013;**2013**:1-14

[9] Scalbert A, Johnson IT, Saltmarsh M. Polyphenols: Antioxidants and beyond. The American Journal of Clinical Nutrition. 2005;**81**:215-217

[10] Sies H. Oxidative stress: From basic research to clinical application. The American Journal of Medicine. 1991;**91**(3 SUPPL. 3):31-38

[11] Jones DP. Redefining Oxidative Stress. Antioxid Redox Signal. 2006;**8**(9-10):1865-1879

[12] Naganathan V. Cardiovascular drugs in older people. Australian Prescriber. 2013;**36**(6):190-194

[13] Deka A, Vita JA. Tea and cardiovascular disease. Pharmacological Research. 2011;**64**:136-145

[14] Lorenzo ADi, Curti V, Tenore GC, Nabavi SM, Daglia M. Effects of tea and coffee consumption on cardiovascular diseases and relative risk factors: An update. Current Pharmaceutical Design. 2017;**23**:2474-2487

[15] Shah S. Primary prevention of cardiovascular disease. InnovAiT Education and Inspiration for General Practice. 2012;**5**(4):195-203

[16] Lloyd-Jones DM, Leip EP, Larson MG, D'Agostino RB, Beiser A, Wilson PWF, et al. Prediction of lifetime risk for cardiovascular disease by risk factor burden at 50 years of age. Circulation. 2006;**113**(6):791-798

[17] Alberti KG, Zimmet P, Shaw J. Metabolic syndrome—A new world wide definition. Lancet. 2005;**366**(9491):1059-1062

[18] Hjemdahl P. Stress and the metabolic syndrome: An interesting but enigmatic association. Circulation. 2002;**106**(21):2634-2636

[19] Barker DJ, Godfrey K, Gluckman P, Harding J, Owens J, Robinson J.

Fetal nutrition and cardiovascular disease in adult life. Lancet. 1993;**341**(8850):938-941

[20] Brunner EJ. Social factors and cardiovascular morbidity. Neuroscience and Biobehavioral Reviews. 2017;**74** (Pt B):260-268

[21] Allgulander C. Anxiety as a risk factor in cardiovascular disease. Current Opinion in Psychiatry. 2016;**29**(1):13-17

[22] Seligman F, Nemeroff CB. The interface of depression and cardiovascular disease: Therapeutic implications. Annals of the New York Academy of Sciences. 2015;**1345**(1):25-35

[23] Smith C, Swart A. *Aspalathus linearis* (rooibos)—A functional food targeting cardiovascular disease. Food & Function. 2018;**9**(10):5041-5058

[24] Schloms L, Swart AC. Rooibos flavonoids inhibit the activity of key adrenal steroidogenic enzymes, modulating steroid hormone levels in H295R cells. Molecules. 2014;**19**(3):3681-3695

[25] Schloms L, Smith C, Storbeck KH, Marnewick JL, Swart P, Swart AC. Rooibos influences glucocorticoid levels and steroid ratios *in vivo* and *in vitro*: A natural approach in the management of stress and metabolic disorders? Molecular Nutrition & Food Research. 2014;**58**(3):537-549

[26] Gathercole LL, Lavery GG, Morgan SA, Cooper MS, Sinclair AJ, Tomlinson JW, et al. 11β-hydroxysteroid dehydrogenase 1: Translational and therapeutic aspects. Endocrine Reviews. 2013;**34**(4):525-555

[27] Schloms L, Storbeck KH, Swart P, Gelderblom WCA, Swart AC. The influence of *Aspalathus linearis* (rooibos) and dihydrochalcones on adrenal steroidogenesis: Quantification of steroid intermediates and end products in H295R cells. The Journal of Steroid Biochemistry and Molecular Biology. 2012;**128**(3-5):128-138

[28] Sobel BE, Schneider DJ. Cardiovascular complications in diabetes mellitus. Current Opinion in Pharmacology. 2005;**5**(2 Spec. Iss):143-148

[29] World Health Organisation. Obesity and Overweight Factsheet No. 311. Geneva: 2018 [cited 2019 Apr 2015]. Available from https://www.who.int/en/news-room/fact-sheets/detail/obesity-and-overweight

[30] Nammi S, Koka S, Chinnala KM, Boini KM. Obesity: An overview on its current perspectives and treatment options. Nutrition Journal. 2004;**3**:1-8

[31] Kruger HS, Venter CS, Vorster HH, THUSA Study. Physical inactivity as a risk factor for cardiovascular disease in communities undergoing rural to urban transition: The THUSA study. Cardiovascular Journal of Southern Africa. 2003;**14**(1):16-23

[32] Eckel RH, Krauss RM. American Heart Association call to action: Obesity as a major risk factor for coronary heart disease. Circulation. 1998;**97**(21):2099-2100

[33] Marnewick JL, Joubert E, Swart P, Van Der Westhuizen F, Gelderblom WC. Modulation of hepatic drug metabolizing enzymes and oxidative status by rooibos (*Aspalathus linearis*) and honeybush (*Cyclopia intermedia*), green and black (*Camellia sinensis*) teas in rats. Journal of Agricultural and Food Chemistry. 2003;**51**(27):8113-8119

[34] Uličná O, Greksák M, Vančová O, Zlatoš L, Galbavý Š, Božek P, et al. Hepatoprotective effect of rooibos tea (*Aspalathus linearis*) on CCl_4-induced liver damage in rats. Physiological Research. 2003;**52**(4):461-466

[35] Sinisalo M, Enkovaara A-L, Kivistö KT. Possible hepatotoxic effect of rooibos tea: A case report. European Journal of Clinical Pharmacology. 2010;**66**(4):427-428

[36] Engels M, Wang C, Matoso A, Maidan E, Wands J. Tea not tincture: Hepatotoxicity associated with rooibos herbal tea. ACG Case Reports Journal. 2013;**1**(1):58-60

[37] Stander MA, Joubert E, De Beer D. Revisiting the caffeine-free status of rooibos and honeybush herbal teas using specific MRM and high resolution LC-MS methods. Journal of Food Composition and Analysis. 2019;**76**:39-43

[38] Cody RB. Ambient profiling of phenolic content in tea infusions by matrix-assisted ionization in vacuum. Journal of the American Society for Mass Spectrometry. 2018;**29**(8):1594-1600

[39] Ferreira D, Kamara BI, Brandt EV, Joubert E. Phenolic compounds from *Cyclopia intermedia* (Honeybush tea). 1. Journal of Agricultural and Food Chemistry. 1998;**46**:3406-3410

[40] Wilson NLW. Cape natural tea products and the U.S. market: Rooibos rebels ready to raid. Applied Economic Perspectives and Policy. 2005;**27**(1):139-148

[41] South African Rooibos Council. Annual Report 2017; 2017. [cited 15 Apr 2019]. Available from: https://sarooibos.co.za/wp/wp-content/uploads/2018/06/2017-SARC-Annual-Report_20180628.pdf

[42] Theron KA. Sensory and phenolic profiling of Cyclopia species (honeybush) and optimisation of the fermentation conditions [Master's thesis]. Stellenbosch: Stellenbosch University; 2012

[43] Joubert E, Richards ES, Van Der Merwe JD, De Beer D, Manley M, Gelderblom WC. Effect of species variation and processing on phenolic composition and *in vitro* antioxidant activity of aqueous extracts of *Cyclopia* spp. (Honeybush tea). Journal of Agricultural and Food Chemistry. 2008;**56**(3):954-963

[44] Joubert E, de Beer D. Rooibos (*Aspalathus linearis*) beyond the farm gate: From herbal tea to potential phytopharmaceutical. South African Journal of Botany. 2011;**77**(4):869-886

[45] Joubert E, Joubert ME, Bester C, de Beer D, De Lange JH. Honeybush (*Cyclopia* spp.): From local cottage industry to global markets—The catalytic and supporting role of research. South African Journal of Botany. 2011;**77**(4):887-907

[46] Manach C, Scalbert A, Morand C, Remesy C, Jimenez L. Polyphenols: Food sources and bioavailability. The American Journal of Clinical Nutrition. 2004;**79**:727-747

[47] Vita JA. Polyphenols and cardiovascular disease: Effects on endothelial and platelet function. The American Journal of Clinical Nutrition. 2005;**81**(1 Suppl):292-297

[48] Stander MA, Van Wyk B-E, Taylor MJC, Long HS. Analysis of phenolic compounds in rooibos tea (*Aspalathus linearis*) with a comparison of flavonoid-based compounds in natural populations of plants from different regions. Journal of Agricultural and Food Chemistry. 2017;**65**(47):10270-10281

[49] Joubert E, Winterton P, Britz TJ, Gelderblom WCA. Antioxidant and pro-oxidant activities of aqueous extracts and crude polyphenolic fractions of rooibos (*Aspalathus linearis*). Journal of Agricultural and Food Chemistry. 2005;**53**:10260-10267

[50] Koeppen BH, Roux DG. Aspalathin: A novel C-glycosyl flavonoid from

Aspalathus linearis. Tetrahedron Letters. 1965;**6**(39):3497-3503

[51] Shimamura N, Miyase T, Umehara K, Warashina T, Fujii S. Phytoestrogens from *Aspalathus linearis*. Biological & Pharmaceutical Bulletin. 2006;**29**(6):1271-1274

[52] Muller CJF, Joubert E, Pheiffer C, Ghoor S, Sanderson M, Chellan N, et al. Z-2-(β-D-glucopyranosyloxy)-3-phenylpropenoic acid, an α-hydroxy acid from rooibos (*Aspalathus linearis*) with hypoglycemic activity. Molecular Nutrition & Food Research. 2013;**57**(12):2216-2222

[53] Magcwebeba TU, Riedel S, Swanevelder S, Swart P, De Beer D, Joubert E, et al. The potential role of polyphenols in the modulation of skin cell viability by *Aspalathus linearis* and *Cyclopia* spp. herbal tea extracts *in vitro*. Journal of Pharmacology. 2016;**68**(11):1440-1453. DOI: 10.1080/14786419.2018.1514398

[54] Augustyn W, Combrinck S, Botha B. Comparison of mangiferin in mango leaf and honeybush infusions. Planta Medica. 2011;**77**(12)

[55] He K, Fan L-L, Wu T-T, Du J. A new xanthone glycoside from *Pyrrosia sheareri*. Natural Product Research. 2018:1-6. https://doi.org/10.1080/14786419.2018.1514398

[56] Young Kim C, Ahn MJ, Kim J. Preparative isolation of mangiferin from *Anemarrhena asphodeloides* rhizomes by centrifugal partition chromatography. Journal of Liquid Chromatography and Related Technologies. 2006;**29**(6):869-875

[57] Stalmach A, Mullen W, Pecorari M, Serafini M, Crozier A. Bioavailability of C-linked dihydrochalcone and flavanone glucosides in humans following ingestion of unfermented and fermented rooibos teas. Journal of Agricultural and Food Chemistry. 2009;**57**(15):7104-7111

[58] Han D, Chen C, Zhang C, Zhang Y, Tang X. Determination of mangiferin in rat plasma by liquid-liquid extraction with UPLC-MS/MS. Journal of Pharmaceutical and Biomedical Analysis. 2010;**51**(1):260-263

[59] Human C, De Beer D, Van Der Rijst M, Aucamp M, Joubert E. Electrospraying as a suitable method for nanoencapsulation of the hydrophilic bioactive dihydrochalcone, aspalathin. Food Chemistry. 2019;**276**:467-474

[60] Villaño D, Pecorari M, Testa MF, Raguzzini A, Stalmach A, Crozier A, et al. Unfermented and fermented rooibos teas (*Aspalathus linearis*) increase plasma total antioxidant capacity in healthy humans. Food Chemistry. 2010;**123**(3):679-683

[61] Canda BD, Oguntibeju OO, Marnewick JL. Effects of consumption of rooibos (*Aspalathus linearis*) and a rooibos-derived commercial supplement on hepatic tissue injury by tert-butyl hydroperoxide in Wistar rats. Oxidative Medicine and Cellular Longevity. 2014;**2014**

[62] Kim S, Chen J, Cheng T, Gindulyte A, He J, He S, et al. PubChem 2019 update: Improved access to chemical data. Nucleic Acids Research. 2018;**47**:1-8

[63] Petrova A, Davids LM, Rautenbach F, Marnewick JL. Photoprotection by honeybush extracts, hesperidin and mangiferin against UVB-induced skin damage in SKH-1 mice. Journal of Photochemistry and Photobiology B: Biology. 2011;**103**(2):126-139

[64] Dludla P, Muller C, Joubert E, Louw J, Essop M, Gabuza K, et al. Aspalathin protects the heart against hyperglycemia-induced oxidative damage by up-regulating Nrf2 expression. Molecules. 2017;**22**(1):129

[65] Halliwell B. Reactive species and antioxidants. Redox biology is a

fundamental theme of aerobic life. Plant Physiology. 2006;**141**(2):312-322

[66] Thannickal V, Fanburg B. Reactive oxygen species in cell signaling. American Journal of Physiology Cell Molecular Physiology. 2000;**279**(6):L1005-L1028

[67] Ma Q. Role of Nrf2 in oxidative stress and toxicity. Annual Review of Pharmacology and Toxicology. 2013;**53**:401-426

[68] Dhalla NS, Temsah RM, Thomas N. Role of oxidative stress in cardiovascular diseases. Journal of Hypertension. 2000;**18**(6):655-673

[69] Ayeleso A, Brooks N, Oguntibeju O. Modulation of antioxidant status in streptozotocin-induced diabetic male Wistar rats following intake of red palm oil and/or rooibos. Asian Pacific Journal of Tropical Medicine. 2014;**7**(7): 536-544

[70] Uličná O, Vančová O, Božek P, Čársky J, Šebeková K, Boor P, et al. Rooibos tea (*Aspalathus linearis*) partially prevents oxidative stress in streptozotocin-induced diabetic rats. Physiological Research. 2006;**55**(2):157-164

[71] Suzuki T, Yamamoto M. Molecular basis of the Keap1-Nrf2 system. Free Radical Biology & Medicine. 2015;**88**(Part B):93-100

[72] Kruidering M, Evan G. Caspase-8 in apoptosis: The beginning of 'the end'? IUBMB Life. 2007;**50**(2):85-90

[73] McLeod CJ, Aziz A, Hoyt RF, McCoy JP, Sack MN. Uncoupling proteins 2 and 3 function in concert to augment tolerance to cardiac ischemia. The Journal of Biological Chemistry. 2005;**280**(39):33470-33476

[74] Johnson R, Dludla PV, Muller CJF, Huisamen B, Essop MF, Louw J. The transcription profile unveils the cardioprotective effect of Aspalathin against lipid toxicity in an *in vitro* H9c2 model. Molecules. 2017;**22**(2):219

[75] Biswas SK. Does the interdependence between oxidative stress and inflammation explain the antioxidant paradox? Oxidative Medicine and Cellular Longevity. 2016;**2016**:17-19

[76] Halliwell B. Are polyphenols antioxidants or pro-oxidants? What do we learn from cell culture and *in vivo* studies? Archives of Biochemistry and Biophysics. 2008;**476**(2):107-112

[77] Marnewick JL, Rautenbach F, Venter I, Neethling H, Blackhurst DM, Wolmarans P, et al. Effects of rooibos (*Aspalathus linearis*) on oxidative stress and biochemical parameters in adults at risk for cardiovascular disease. Journal of Ethnopharmacology. 2011;**133**(1):46-52

[78] Apak R, Gorinstein S, Böhm V, Schaich KM, Özyürek M, Güçlü K. Methods of measurement and evaluation of natural antioxidant capacity/activity (IUPAC technical report)*. Pure and Applied Chemistry. 2013;**85**(5):957-998

[79] Kamara BI, Brandt EV, Ferreira D, Joubert E. Polyphenols from Honeybush tea (*Cyclopia intermedia*). Journal of Agricultural and Food Chemistry. 2003;**51**(13):3874-3879

[80] Joubert E, Gelderblom WCA, Louw A, de Beer D. South African herbal teas: *Aspalathus linearis*, *Cyclopia* spp. and Athrixia phylicoides—A review. Journal of Ethnopharmacology. 2008;**119**(3):376-412

[81] Saha S, Sadhukhan P, Sinha K, Agarwal N, Sil PC. Mangiferin attenuates oxidative stress induced renal cell damage through activation of PI3K induced Akt and Nrf-2 mediated

signaling pathways. Biochemistry and Biophysics Reports. 2016;**5**:313-327

[82] Suchal K, Malik S, Khan SI, Malhotra RK, Goyal SN, Bhatia J, et al. Protective effect of mangiferin on myocardial ischemia-reperfusion injury in streptozotocin-induced diabetic rats: Role of AGE-RAGE/MAPK pathways. Scientific Reports. 2017;**7**(January):1-11

[83] Mahali SK, Verma N, Manna SK. Advanced glycation end products induce lipogenesis: Regulation by natural xanthone through inhibition of ERK and NF-κB. Journal of Cellular Physiology. 2014;**229**(12):1972-1980

[84] Sesso HD, Buring JE, Christen WG, Kurth T, Belanger C, MacFadyen J, et al. Vitamins E and C in the prevention of cardiovascular disease in men: The physicians' health study II randomized controlled trial. Journal of the American Medical Association. 2008;**300**(18):2123-2133

[85] Manson JE, Cook NR, Lee I-M, Christen W, Bassuk SS, Mora S, et al. Vitamin D supplements and prevention of cancer and cardiovascular disease. The New England Journal of Medicine. 2019;**380**(1):33-44

[86] Lee D-H, Folsom AR, Harnack L, Halliwell B, Jacobs DR. Does supplemental vitamin C increase cardiovascular disease risk in women with diabetes? The American Journal of Clinical Nutrition. 2004;**80**(5):1194-1200

[87] Miller ER, Pastor-Barriuso R, Dalal D, Riemersma RA, Appel LJGE. Review meta-analysis: High-dosage vitamin E supplementation may increase. Annals of Internal Medicine. 2005;**142**(1):37-46

[88] Rapola JM, Virtamo J, Ripatti S, Huttunen JK, Albanes D, Taylor PR, et al. Randomised trial of α-tocopherol and β-carotene supplements on incidence of major coronary events in men with previous myocardial infarction. Lancet. 1997;**349**(9067):1715-1720

[89] Snijman PW, Joubert E, Ferreira D, Li XC, Ding Y, Green IR, et al. Antioxidant activity of the dihydrochalcones aspalathin and nothofagin and their corresponding flavones in relation to other rooibos (*Aspalathus linearis*) flavonoids, epigallocatechin gallate, and Trolox. Journal of Agricultural and Food Chemistry. 2009;**57**(15):6678-6684

[90] Murakami S, Miura Y, Hattori M, Matsuda H, Malherbe CJ, Muller CJF, et al. Cyclopia extracts enhance Th1-, Th2-, and Th17-type T cell responses and induce Foxp3^{+} cells in murine cell culture. Planta Medica. 2018;**84**(5):311-319

[91] Yang S, Lee C, Lee BS, Park EK, Kim KM, Bae JS. Renal protective effects of aspalathin and nothofagin from rooibos (*Aspalathus linearis*) in a mouse model of sepsis. Pharmacological Reports. 2018;**70**(6):1195-1201

[92] Thangaraj K, Vaiyapuri M. Orientin, a C-glycosyl dietary flavone, suppresses colonic cell proliferation and mitigates NF-κB mediated inflammatory response in 1,2-dimethylhydrazine induced colorectal carcinogenesis. Biomedicine & Pharmacotherapy. 2017;**96**:1253-1266

[93] Magcwebeba T, Swart P, Swanevelder S, Joubert E, Gelderblom W. Anti-inflammatory effects of *Aspalathus linearis* and *Cyclopia* spp. Extracts in a UVB/Keratinocyte (HaCaT) model utilising interleukin-1-Accumulation as biomarker. Molecules. 2016;**21**(10):1-21

[94] Im AR, Yeon SH, Lee JS, Um KA, Ahn YJ, Chae S. Protective effect of fermented *Cyclopia intermedia* against UVB-induced damage in HaCaT human

keratinocytes. BMC Complementary and Alternative Medicine. 2016;**16**(1):261

[95] Lee W, Bae JS. Anti-inflammatory effects of aspalathin and nothofagin from rooibos (*Aspalathus linearis*) *in vitro* and *in vivo*. Inflammation. 2015;**38**(4):1502-1516

[96] Lee W, Ku S-K, Bae J-S. Ameliorative Effect of Vicenin-2 and Scolymoside on TGFBIp-Induced Septic Responses. Inflammation. 2015;**38**(6):2166-2177

[97] Ku S-K, Bae J-S. Vicenin-2 and scolymoside inhibit high-glucose-induced vascular inflammation *in vitro* and *in vivo*. Canadian Journal of Physiology and Pharmacology. 2016;**94**(3):287-295

[98] Hendricks R, Pool EJ. The *in vitro* effects of rooibos and black tea on immune pathways. Journal of Immunoassay and Immunochemistry. 2010;**31**(2):169-180

[99] Mueller M, Hobiger S, Jungbauer A. Anti-inflammatory activity of extracts from fruits, herbs and spices. Food Chemistry. 2010;**122**(4):987-996

[100] Baba H, Ohtsuka Y, Haruna H, Lee T, Nagata S, Maeda M, et al. Studies of anti-inflammatory effects of rooibos tea in rats. Pediatrics International. 2009;**51**(5):700-704

[101] Kunishiro K, Tai A, Yamammoto I. Effects of rooibos tea extract on antigen-specific antibody production and cytokine generation *in vitro* and *in vivo*. Bioscience, Biotechnology, and Biochemistry. 2001;**65**(10):2137-2145

[102] Mazibuko-Mbeje SE, Dludla PV, Roux C, Johnson R, Ghoor S, Joubert E, et al. Aspalathin-enriched green rooibos extract reduces hepatic insulin resistance by modulating PI3K/Akt and AMPK pathways. International Journal of Molecular Sciences. 2019;**20**(3):633

[103] Smit SE, Johnson R, Van Vuuren MA, Huisamen B. Myocardial glucose clearance by Aspalathin treatment in Young, mature, and obese insulin-resistant rats. Planta Medica. 2018;**84**(2):75-82

[104] Mazibuko SE, Joubert E, Johnson R, Louw J, Opoku AR, Muller CJF. Aspalathin improves glucose and lipid metabolism in 3T3-L1 adipocytes exposed to palmitate. Molecular Nutrition & Food Research. 2015;**59**(11):2199-2208

[105] Sanderson M, Mazibuko SE, Joubert E, De Beer D, Johnson R, Pheiffer C, et al. Effects of fermented rooibos (*Aspalathus linearis*) on adipocyte differentiation. Phytomedicine. 2014;**21**:109-117

[106] Pheiffer C, Dudhia Z, Louw J, Muller C, Joubert E. *Cyclopia maculata* (honeybush tea) stimulates lipolysis in 3T3-L1 adipocytes. Phytomedicine. 2013;**20**(13):1168-1171

[107] Dudhia Z, Louw J, Muller C, Joubert E, De Beer D, Kinnear C, et al. *Cyclopia maculata and Cyclopia subternata* (honeybush tea) inhibits adipogenesis in 3T3-L1 pre-adipocytes. Phytomedicine. 2013;**20**(5):401-408

[108] Beltrán-Debón R, Rull A, Rodríguez-Sanabria F, Iswaldi I, Herranz-López M, Aragonès G, et al. Continuous administration of polyphenols from aqueous rooibos (*Aspalathus linearis*) extract ameliorates dietary-induced metabolic disturbances in hyperlipidemic mice. Phytomedicine. 2011;**18**(5):414-424

[109] Johnson R, Dludla P, Joubert E, February F, Mazibuko S, Ghoor S, et al. Aspalathin, a dihydrochalcone C-glucoside, protects H9c2 cardiomyocytes against high glucose induced shifts in substrate preference and apoptosis. Molecular Nutrition & Food Research. 2016;**60**(4):922-934

[110] Schulze AE, De Beer D, Mazibuko SE, Muller CJF, Roux C, Willenburg EL, et al. Assessing similarity analysis of chromatographic fingerprints of Cyclopia subternata extracts as potential screening tool for *in vitro* glucose utilisation. Analytical and Bioanalytical Chemistry. 2016;**408**(2):639-649

[111] Kamakura R, Son MJ, de Beer D, Joubert E, Miura Y, Yagasaki K. Antidiabetic effect of green rooibos (*Aspalathus linearis*) extract in cultured cells and type 2 diabetic model KK-ay mice. Cytotechnology. 2015;**67**(4):699-710

[112] Ku SK, Kwak S, Kim Y, Bae JS. Aspalathin and nothofagin from rooibos (*Aspalathus linearis*) inhibits high glucose-induced inflammation *in vitro* and *in vivo*. Inflammation. 2014;**38**(1):445-455

[113] Dludla PV, Muller CJF, Louw J, Joubert E, Salie R, Opoku AR, et al. The cardioprotective effect of an aqueous extract of fermented rooibos (*Aspalathus linearis*) on cultured cardiomyocytes derived from diabetic rats. Phytomedicine. 2014;**21**(5):595-601

[114] Mathijs I, da Cunha DA, Himpe E, Ladriere L, Chellan N, Roux CR, et al. Phenylpropenoic acid glucoside augments pancreatic beta cell mass in high-fat diet-fed mice and protects beta cells from ER stress-induced apoptosis. Molecular Nutrition & Food Research. 2014;**58**(10):1980-1990

[115] Saleh S, El-Maraghy N, Reda E, Barakat W. Modulation of diabetes and dyslipidemia in diabetic insulin-resistant rats by mangiferin: Role of adiponectin and TNF-α. Anais da Academia Brasileira de Ciências. 2014;**86**(4):1935-1948

[116] Chellan N, Joubert E, Strijdom H, Roux C, Louw J, Muller CJF. Aqueous extract of unfermented honeybush (*Cyclopia maculata*) attenuates stz-induced diabetes and β-cell cytotoxicity. Planta Medica. 2014;**80**(8-9):622-629

[117] Son MJ, Minakawa M, Miura Y, Yagasaki K. Aspalathin improves hyperglycemia and glucose intolerance in obese diabetic ob/ob mice. European Journal of Nutrition. 2013;**52**(6):1607-1619

[118] Mazibuko SE, Muller CJF, Joubert E, De Beer D, Johnson R, Opoku AR, et al. Amelioration of palmitate-induced insulin resistance in C2C12 muscle cells by rooibos (*Aspalathus linearis*). Phytomedicine. 2013;**20**(10):813-819

[119] Muller CJF, Joubert E, de Beer D, Sanderson M, Malherbe CJ, Fey SJ, et al. Acute assessment of an aspalathin-enriched green rooibos (*Aspalathus linearis*) extract with hypoglycemic potential. Phytomedicine. 2012;**20**(1):32-39

[120] Muller CJF, Joubert E, Gabuza K, De Beer D, Fey SJ, Louw J. Assessment of the antidiabetic potential of an aqueous extract of Honeybush (*Cyclopia intermedia*) in streptozotocin and obese insulin resistant Wistar rats. In: Rasooli I, editor. Phytochemistry-Bioactivities and Impact on Health. London: IntechOpen; 2011:313-332. DOI: 10.5772/28574

[121] Kawano A, Nakamura H, Hata S, Minakawa M, Miura Y, Yagasaki K. Hypoglycemic effect of aspalathin, a rooibos tea component from *Aspalathus linearis*, in type 2 diabetic model db/db mice. Phytomedicine. 2009;**16**(5):437-443

[122] Sun D, Huang J, Zhang Z, Gao H, Li J, Shen M, et al. Luteolin limits infarct size and improves cardiac function after myocardium ischemia/reperfusion injury in diabetic rats. PLoS One. 2012;**7**(3)

[123] Wu X, Xu T, Li D, Zhu S, Chen Q, Hu W, et al. ERK/PP1a/PLB/SERCA2a

and JNK pathways are involved in luteolin-mediated protection of rat hearts and cardiomyocytes following ischemia/reperfusion. PLoS One. 2013;**8**(12)

[124] Persson IA-L, Persson K, Hägg S, Andersson RGG. Effects of green tea, black tea and rooibos tea on angiotensin-converting enzyme and nitric oxide in healthy volunteers. Public Health Nutrition. 2010;**13**(05):730

[125] Persson IA-L, Josefsson M, Persson K, Andersson RGG. Tea flavanols inhibit angiotensin-converting enzyme activity and increase nitric oxide production in human endothelial cells. The Journal of Pharmacy and Pharmacology. 2006;**58**(8):1139-1144

[126] Persson IA-L. The pharmacological mechanism of angiotensin-converting enzyme inhibition by green tea, rooibos and enalaprilat—A study on enzyme kinetics. Phytother Research. 2012;**26**(4):517-521

[127] Fu XC, Wang MW, Li SP, Wang HL. Anti-apoptotic effect and the mechanism of orientin on ischaemic/reperfused myocardium. Journal of Asian Natural Products Research. 2006;**8**(3):265-272

[128] Lu N, Sun Y, Zheng X. Orientin-induced cardioprotection against reperfusion is associated with attenuation of mitochondrial permeability transition. Planta Medica. 2011;**77**(10):984-991

[129] Pantsi WG, Marnewick JL, Esterhuyse AJ, Rautenbach F, Van Rooyen J. Rooibos (*Aspalathus linearis*) offers cardiac protection against ischaemia/reperfusion in the isolated perfused rat heart. Phytomedicine. 2011;**18**(14):1220-1228

[130] Khan AU, Gilani AH. Selective bronchodilatory effect of rooibos tea (*Aspalathus linearis*) and its flavonoid, chrysoeriol. European Journal of Nutrition. 2006;**45**(8):463-469

[131] Guerrero L, Castillo J, Quiñones M, Garcia-Vallvé S, Arola L, Pujadas G, et al. Inhibition of angiotensin-converting enzyme activity by flavonoids: Structure-activity relationship studies. PLoS One. 2012;**7**(11):1-11

[132] Inoguchi T, Li P, Umeda F, Yu HY, Kakimoto M, Imamura M, et al. High glucose level and free fatty acid stimulate reactive oxygen species production through protein kinase-C dependent activation of NAD(P)H oxidase in cultured vascular cells. Diabetes. 2000;**49**:1939-1945

[133] Esser N, Legrand-Poels S, Piette J, Scheen AJ, Paquot N. Inflammation as a link between obesity, metabolic syndrome and type 2 diabetes. Diabetes Research and Clinical Practice. 2014;**105**(2):141-150

[134] Reaven GM. Pathophysiology of insulin resistance in human disease. Physiological Reviews. 1995;**75**(3):473-486

[135] Sookoian S, Pirola CJ. Review article: Shared disease mechanisms between non-alcoholic fatty liver disease and metabolic syndrome—Translating knowledge from systems biology to the bedside. Alimentary Pharmacology & Therapeutics. 2019;**49**(November 2018):516-527

[136] Huang X, Liu G, Guo J, Su Z. The PI3K/Akt pathway in obesity and type 2 diabetes. International Journal of Biological Sciences. 2018;**14**(11):1483-1496

[137] Lee EM, Lee SS, Chung BY, Cho JY, Lee IC, Ahn SR, et al. Pancreatic lipase inhibition by C-glycosidic flavones isolated from *Eremochloa* ophiuroides. Molecules. 2010;**15**(11):8251-8259

[138] Krebs DL. SOCS Proteins: Negative Regulators of Cytokine Signaling. Stem Cells. 2001;**19**:378-387

[139] Jack BU, Malherbe CJ, Huisamen B, Gabuza K, Mazibuko-Mbeje S, Schulze AE, et al. A polyphenol-enriched fraction of *Cyclopia intermedia* decreases lipid content in 3T3-L1 adipocytes and reduces body weight gain of obese db/db mice. South African Journal of Botany. 2017;**110**:216-229

[140] Tansey JT, Sztalryd C, Hlavin EM, Kimmel AR, Londos C. The central role of perilipin a in lipid metabolism and adipocyte lipolysis. IUBMB Life. 2004;**56**(7):379-385

[141] Poher AL, Altirriba J, Veyrat-Durebex C, Rohner-Jeanrenaud F. Brown adipose tissue activity as a target for the treatment of obesity/insulin resistance. Frontiers in Physiology. 2015;**6**(4):1-9

[142] Schrauwen P, Hesselink MKC. The role of uncoupling protein 3 in fatty acid metabolism: Protection against lipotoxicity? The Proceedings of the Nutrition Society. 2004;**63**(2):287-292

[143] Na L, Zhang Q, Jiang S, Du S, Zhang W, Li Y, et al. Mangiferin supplementation improves serum lipid profiles in overweight patients with hyperlipidemia: A double-blind randomized controlled trial. Scientific Reports. 2015;**5**:1-9

[144] Laakso M. Hyperglycemia and cardiovascular disease in type 2 diabetes. Diabetes. 1999;**48**(5):937

[145] Maritim AC, Sanders RA, Watkins JB. Diabetes, oxidative stress, and antioxidants: A review. Journal of Biochemical and Molecular Toxicology. 2003;**17**(1):24-38

[146] Sasaki M, Nishida N, Shimada MA. Beneficial role of rooibos in diabetes mellitus: A systematic review and meta-analysis. Molecules. 2018;**23**(4):839

[147] Himpe E, Cunha DA, Song I, Bugliani M, Marchetti P, Cnop M, et al. Phenylpropenoic acid glucoside from rooibos protects pancreatic Beta cells against cell death induced by acute injury. PLoS One. 2016;**11**(6):e0157604

[148] Song I, Roels S, Martens GA, Bouwens L. Circulating microRNA-375 as biomarker of pancreatic beta cell death and protection of beta cell mass by cytoprotective compounds. PLoS One. 2017;**12**(10):e0186480

[149] Beelders T, Brand DJ, de Beer D, Malherbe CJ, Mazibuko SE, Muller CJF, et al. Benzophenone C- and O-glucosides from *Cyclopia genistoides* (Honeybush) inhibit mammalian α-glucosidase. Journal of Natural Products. 2014;**77**(12):2694-2699

[150] Liu W, Wang H, Meng F. In silico modeling of aspalathin and nothofagin against SGLT2. Journal of Theoretical and Computational Chemistry. 2015;**14**(08):1550056

[151] Chen D, Chen R, Wang R, Li J, Xie K, Bian C, et al. Probing the catalytic promiscuity of a regio- and stereospecific C-glycosyltransferase from *Mangifera indica*. Angewandte Chemie, International Edition. 2015;**54**(43):12678-12682

[152] Santos CMM, Freitas M, Fernandes E. A comprehensive review on xanthone derivatives as α-glucosidase inhibitors. European Journal of Medicinal Chemistry. 2018;**157**:1460-1479

[153] Jesus AR, Vila-Viçosa D, Machuqueiro M, Marques AP, Dore TM, Rauter AP. Targeting type 2 diabetes with *C*-glucosyl dihydrochalcones as selective sodium glucose Co-transporter 2 (SGLT2) inhibitors: Synthesis and biological evaluation. Journal of Medicinal Chemistry. 2017;**60**(2):568-579

[154] Ehrenkranz JRL, Lewis NG, Ronald Kahn C, Roth J. Phlorizin: A review. Diabetes/Metabolism Research and Reviews. 2005;**21**(1):31-38

[155] Chellan N. Aqueous extract of unfermented honeybush (*Cyclopia maculata*) attenuates STZ-Induced diabetes and β-cell cytotoxicity. Planta Medica. 2014;80:622-629

[156] Singh AK, Raj V, Keshari AK, Rai A, Kumar P, Rawat A, et al. Isolated mangiferin and naringenin exert antidiabetic effect via PPARγ/GLUT4 dual agonistic action with strong metabolic regulation. Chemico-Biological Interactions. 2018;**280**:33-44

[157] Miura T, Ichiki H, Hashimoto I, Iwamoto N, Kao M, Kubo M, et al. Antidiabetic activity of a xanthone compound, mangiferin. Phytomedicine. 2001;**8**(2):85-87

[158] Schulman IH, Zhou MS, Raij L. Nitric oxide, angiotensin II, and reactive oxygen species in hypertension and atherogenesis. Current Hypertension Reports. 2005;**7**(1):61-67

[159] Hadi HAR, Carr CS, Al Suwaidi J. Endothelial dysfunction: Cardiovascular risk factors, therapy, and outcome. Vascular Health and Risk Management. 2005;**1**(3):183-198

[160] Gilani AH, Khan A, Ghayur MN, Ali SF, Herzig JW. Antispasmodic effects of rooibos tea (*Aspalathus linearis*) is mediated predominantly through K^+ channel activation. Basic Clinical Pharmacology Toxicology. 2006;**99**(5):365-373

[161] Maqeda Z. Investigating the Modulating Effects of Afriplex GRTTM Extract on Vascular Function and Antioxidant Status in Obese Wistar Rats [Master's thesis]. Stellenbosch: Stellenbosch University; 2018

[162] Smit-van Schalkwyk M. Rooibos and Melatonin: Putative Modulation of Nicotine-induced Effects on Vascular Function [Doctoral thesis]. Stellenbosch: Stellenbosch University; 2016

[163] Williams SM, Haines JL, Moore JH. The use of animal models in the study of complex disease: All else is never equal or why do so many human studies fail to replicate animal findings? BioEssays. 2004;**26**(2):170-179

Chapter 4

Probiotics and Other Bioactive Compounds with Proven Effect against Obesity and Hypertension: Food Design Opportunities from Lulo Fruit (*Solanum quitoense*)

Noelia Betoret, Leidy Indira Hinestroza, Lucía Seguí and Cristina Barrera

Abstract

This book chapter aims to identify those bioactive compounds that are the most effective in obesity and hypertension prevention and/or treatment, these being the two main disorders associated with metabolic syndrome. Focusing on probiotics and phytochemicals, the document will provide evidences from both in vitro and in vivo studies as well as information about the action mechanisms and how they are affected by the interaction with other food ingredients, the food matrix in which they are placed, etc. Given its high antioxidant capacity, in part due to its spermidine content, lulo fruit has generated considerable interest among health researchers. This, together with its exotic organoleptic properties, offers interesting growth opportunities for the design of new food products from lulo fruit. This book chapter will also discuss some of them.

Keywords: probiotics, phytochemicals, metabolic syndrome, spermidine, lulo

1. Foods, technology, and metabolic syndrome

Overweight and obesity are defined as "abnormal or excessive fat accumulation that may impair health" [1]. Since 1975, obesity has almost tripled worldwide so that in 2016, 39% of the adult population and 18% of children and adolescents were overweight. Very often, a high body mass index (BMI ≥ 25–30) is associated with other metabolic abnormalities, such as high blood pressure (hypertension), high blood sugar (hyperglycemia), high serum triglycerides, and low serum high-density lipoprotein (HDL) [2]. The occurrence of at least three of these interconnected physiological, biochemical, clinical, and metabolic factors that directly increase the risk of atherosclerotic cardiovascular disease, cancer, and type 2 diabetes is known as metabolic syndrome [3]. The International Diabetes Federation estimates that one-quarter of the world's adult population suffers from this syndrome, with little

IntechOpen

difference between developed and developing countries. Main factors contributing to it include, beyond the genetic susceptibility, the increased consumption of calorie-dense food and the scarce physical activity. Given that metabolic syndrome can occur in several forms, according to the combination of the different components, it is apparently difficult to treat it pharmacologically, being lifestyle change the most effective preventive approach. However, the fact that a low-grade chronic inflammatory state accompanies the metabolic syndrome in any of its forms suggests that anti-inflammatory therapies could have a place in its prevention and treatment [4].

Inflammation is a response of the body's immune system to harmful stimuli. In the case of metabolic syndrome, inflammation takes place in response to imbalance of blood glucose and insulin levels or insulin resistance, which leads to unhealthy high concentration of unused sugar in the bloodstream that is sent to the liver, muscle, or pancreas [5, 6]. Once there, the sugar is converted into fat, thus leading to progressive adipocyte enlargement. Hypertrophy reduces blood supply to adipocytes and causes hypoxia. Subsequent necrosis and macrophage infiltration into adipose tissue lead to overproduction of reactive oxygen species (ROS), low-density lipoproteins (LDLs), inflammatory cytokines (tumor necrosis factor-alpha, interleukin-6, adiponectin, etc.), and C-reactive protein (CRP). High-fat diets, frequently consumed by obese individuals, aggravate this problem both directly, when the fat is rich in saturated fatty acids, and indirectly, through effects on the microbiota and intestinal permeability [5]. The normal blood plasma concentration of CRP varies between 0.08 and 0.3 mg/dL in healthy adults and reaches values between 2 and 10 mg/dL in individuals suffering from metabolic syndrome. High levels of the inflammatory marker CRP in blood are associated with increased odds of having plaque in the carotid arteries and, therefore, with increased risk of myocardial infarction and stroke [7].

In addition to CRP, some dietary food components involved as intermediates in various metabolic pathways appear altered in population with metabolic syndrome, making possible its use as biomarkers. Polyunsaturated fatty acids (PUFA), specifically eicosapentaenoic acid (20:5 n-3; EPA) and docosahexaenoic acid (22:6 n-3; DHA), are inversely related to metabolic syndrome in adults [8]; several studies have stablished an association between this syndrome and selenium blood concentration [9] or serum levels of vitamin B12 [10]; Urrunaga-Pastor et al. [11] also found an association between vitamin D deficiency and hyperinsulinemia; adults with metabolic syndrome also have suboptimal concentrations of several antioxidants (retinyl esters, vitamin C, and all carotenoid concentrations, except lycopene), partly due to the lower intake of fruit and vegetables by these individuals [12]. These results reinforce the relationship between diet and the incidence of metabolic syndrome.

Evidence from prospective observational in vitro and in vivo studies (preclinical and clinical trials) has converged to support the importance of individual nutrient or food intake and dietary patterns in the prevention and management of obesity and metabolic syndrome. In vitro studies seek to determine the biochemical mechanisms at the cellular level as physiological ones, which are involved in the proper functioning both at the transcriptional and protein expression in the pancreas, skeletal muscle, liver, and adipose tissue. In vivo studies allow establishing a cause-effect relationship in experimental animals (preclinical trials) or in humans (clinical trials). While in vitro or animal in vivo studies are standardized and there are countless works done, even today a standard profile for clinical trials in humans with metabolic syndrome has not yet been established. Clinical trials about therapeutic efficacy for metabolic syndrome are scarce and concentrated in the last 8 years in high-income countries (USA, Italy, and Spain). Interventions that affect three or more factors and evaluate various outcome variables are reduced, highlighting the lifestyle factors (diet and physical activity) as the most important in this multifactorial syndrome [13]. Specifically, a low intake of saturated and total fat; reduced consumption of sodium, simple sugars, and high glycemic index foods; and increased intake of fruits, vegetables, legumes, and whole grains are suggested to be the most effective actions in reducing the incidence of obesity and cardiovascular disease. However, the urban lifestyle leads us to mainly consume foods in processed form, which reinforces the decisive role that the food industry plays in the promotion of healthy diets. In fact, in recent years the supply of functional foods with a reduced content of fat, sugar or salt, as well as that of functional foods formulated with phytosterols or polyunsaturated fatty acids has increased considerably. In order to achieve this, not only traditional techniques of food formulation and blending or cultivation and breeding are involved but also more recent ones, such as microencapsulation, vacuum impregnation, or coating with edible films [14]. Moreover, the increasing knowledge about the negative impact that processing and cooking techniques have on the concentration and functionality of the active compounds naturally present in foods has encouraged the use of alternative techniques, such as the application of high-pressure homogenization replacing the pasteurization or freeze-drying instead of hot air drying.

Later in this book chapter, the most relevant bioactive compounds with proven effect against any disorder associated with the metabolic syndrome are listed. Focusing on probiotic microorganisms and phytochemical compounds, evidences obtained from in vitro, in vivo, or clinical studies in the last 10 years have been compiled. Finally, new functional foods made from lulo fruit are suggested as being suitable for metabolic syndrome prevention and/or amelioration.

2. Food components against metabolic syndrome

Main food components considered in the literature as having potential ameliorating effect on any disorder associated with metabolic syndrome may be included in one of the following groups:

- *Fiber*, both soluble (e.g., pectins, beta-glucans, naturally occurring gums, inulin, psyllium) and insoluble (e.g., cellulose, hemicellulose, lignin), is reported to have laxative properties and to mitigate both hypercholesterolemia and hyperglycemia [15]. When fermented by probiotic bacteria, *prebiotic fiber* (fiber that resists digestion in the stomach and the small intestine and reaches the colon intact) breaks down into short-chain fatty acids (butyrate, acetate, and propionate), which are reported to enhance glucose and fat metabolism [16].

- *Monounsaturated fats, polyunsaturated fats (both omega-3 and omega-6), plant sterols, and essential fatty acids* instead of saturated fats and trans-fatty acids are proved to be effective in decreasing total cholesterol and increasing the blood level of high-density lipoproteins (HDLs) [17].

- *Vitamin E and C* consumption is associated to a reduction of vascular risk by decreasing oxidative stress (lipid peroxidation) and proinflammatory cytokines [17]. Improved vitamin C status is hypothesized to alleviate endotoxemia and its consequent proinflammatory responses that are suggested to initiate insulin resistance and related metabolic disorders; on the contrary, inadequate vitamin C status contributes to small intestinal bacterial overgrowth, transcytosis of enteric bacteria, and an elevation of circulating lipopolysaccharide, which elicits a low-grade inflammatory response [18]. As for *vitamin D*, it reduces the intestinal absorption of fat by increasing that of calcium [17].

- *Bioactive peptides*, having a size range of 2–50 amino acids, have potential to regulate blood pressure and glycemia, reduce cholesterol level and body mass, and scavenge free radicals [19]. Lactotripeptides isoleucine-proline-proline and valine-proline-proline, whose concentrations increase during the ripening process of cheese, are particularly considered as strong antihypertensive agents [20]. In the case of obesity, foods that contain bioactive peptides provide a satiating effect and lead to appetite suppression.

- *Minerals'* (selenium, magnesium, and zinc) ability to decline metabolic syndrome is related with their antioxidant properties and their participation in insulin synthesis and regulation [20]. Moreover, calcium intake (1200 mg/day) has been demonstrated to increase fat mass loss in overweight and obese adults [21]. Mechanisms to explain this effect include that during low calcium intake, more calcium enters adipose tissues cells and subsequently stimulates the expression of lipogenic genes in parallel with suppressing lipolysis [22]. Also, calcium increases fecal fat loss by binding to fat in the lumen and forming non-absorbed complexes [23]. This ability of calcium to decrease lipogenesis may be enhanced due to a synergistic effect with other components in dairy products (vitamin D and angiotensin-converting enzyme inhibitors) [21].

- *Essential amino acids*, mainly histidine and glycine, are associated with a decrease in insulin resistance and blood pressure, respectively, although conclusive evidence is lacking and additional studies are needed [17].

- *Phytochemicals*, mainly polyphenols (phenolic acids, curcuminoids, stilbenes, lignans, flavonoids, flavonols, flavones, anthocyanins, etc.) but also other bioactive components present in small quantities in fruits and vegetables (triterpenes, carotenoids, etc.), have demonstrated anti-inflammatory, anti-oxidative, antiadiposity, and cardioprotective functions in a huge amount of studies [24–28]. Some phytochemicals, among which are caffeine, ephedrine, capsaicin, and salicylic acid, also act as thermogenic compounds that produce heat from lipids and fats, thus burning extra calories and preventing the accumulation of fat in body tissues [20].

- *Probiotics* are microorganisms that improve the availability and digestibility of nutrients while maintaining the balance of intestinal microflora in the gut [20]. Mainly belonging to the *Lactobacillus* and *Bifidobacterium* genera, probiotics emerge as prospective biotherapies in the management of metabolic disorders including obesity and diabetes by counteracting the adverse effects of a high-fat diet [29].

2.1 Phytochemicals

Phytochemicals are naturally occurring plant chemicals that, beyond providing plants with color, odor, and flavor, can influence chemical processes within human bodies in a beneficial way. Main phytochemicals under research include carotenoids (β-carotene, lycopene, lutein, zeaxanthin) and polyphenols, which include phenolic acids, flavonoids, and stilbenes/lignans [30]. Flavonoids can be further divided into groups based on their similar chemical structure, such as anthocyanins, flavones, flavanones, isoflavones, flavonols, and flavanols. Flavanols further are classified as catechins, epicatechins, and proanthocyanidins.

Phytochemical compounds have gained popularity in recent years due to their broadly documented effect in cancer prevention among other biological effects, such as the prevention and treatment of obesity, cholesterol, and diabetes [31]. Mechanisms employed by phytochemicals in ameliorating metabolic syndrome risk factors are diverse and dependent on their particular chemical structure. Whereas catechins mainly induce fat oxidation and improve endothelial function, cyanidins and theaflavins inhibit enzymes involved in the synthesis of fatty acids and triglycerides. Other phytochemicals, such as gallic acid, quercetin, and capsaicin, reduce preadipocyte proliferation by induction of cell apoptosis, while low-molecular proanthocyanidins have the ability to inhibit the activity of specific angiotensin-converting enzyme. Having a similar chemical structure, isoflavones can also influence the activity of human estrogens.

For some bioactive components, several in vitro and in vivo (either animals or humans) studies have been performed; however, as evidenced in **Tables 1–3**, results differ among them. While in vitro or animal studies usually yield positive results, clinical human studies are still inconclusive. The lack of standardization or aspects related to the dose or duration of supplementation may be the cause of these results. Moreover, studies both in vitro and with animals (**Tables 1** and **2**) have been carried out with synthetic components (only spermidine was obtained directly from lulo fruit), while in vivo studies with humans have been carried out mainly with extracts including the bioactive compounds (**Table 3**).

Phytochemical(s)	Methodology	Beneficial effect(s)	Reference
Synthesis naringenin	3T3-L1 cell line and *Pemphigus vulgaris*-treated HaCaT cell line. Naringenin was added to the cell growing media in a dose of 25 μg/mL, and effect was measured after 24, 48, 72, 96, or 120 hours	Anti-adipogenic, antioxidant, anti-inflammatory, and antiapoptosis effects	[32, 33]
Pentacyclic triterpenes (oleanolic acid, 18β-glycyrrhetinic acid, ursolic acid, celastrol, maslinic acid, ilexgenin A)	3 T3-L1 cell line or 10 T1/2 cells and primary fat SVF cells or HepG2 cells Doses and times were not specified	Decreased obesity-induced inflammation, stimulated lipolysis, and decreased adipocyte differentiation	[26]
Synthesis carvacrol	Cyclooxygenase-2 assay IC50 = 0.8 μM	Anti-inflammatory potential	[34]
Spermidine (from ethanolic lulo pulp extract)	In vitro measurement of angiotensin-converting enzyme inhibition IC50 = 1.8 ppm	Hypertension control	[35]
Flavonoids (naringenin, rutin, hesperidin, resveratrol, naringin, and quercetin)	3 T3-L1 cell line IC50 = 40.4 μM for quercetin IC50 ≥ 500 μM for naringenin, rutin, hesperidin, resveratrol, and naringin	Quercetin efficiently inhibited cell population growth and increased induction of apoptosis	[36]
Resveratrol	Maturing preadipocytes and adipocytes. Dose not specified	Decreased adipogenesis, increased lipolysis, induced apoptosis, and reduced lipogenesis and proliferation, thereby contributing to reduce lipid accumulation. Reduced inflammatory response and improved insulin sensitivity	[37]

Phytochemical(s)	Methodology	Beneficial effect(s)	Reference
Ajoene	Mature 3T3-L1 adipocytes Ajoene at 200 μM decreased cell viability in 50% after 24 hours of treatment	Influenced the regulation of fat cell number	[38]
Green tea catechins	3 T3-L1 cells Epigallocatechin or epigallocatechin gallate was added to the cell growing media in a dose from 1, 10, 50, 100, 200, and 200 μM for some hours until some days (8–16 days)	Increased apoptosis and decreased preadipocyte proliferation	[27]

Table 1.
Fruits, vegetables (or extracts), and phytochemicals endorsed by recent in vitro studies.

Phytochemical(s)	Methodology	Beneficial effect(s)	Reference
Naringenin	Rats and mice 10 mg/kg•day by oral gavage for 4 weeks; 0.1% in an experimental diet for 6 months; 1% and 3% in a high-fat diet for 4 and 30 weeks, respectively	Antioxidant, antihyperlipidemic, anti-obesity, antihyperglycemic, anti-diabetic, anti-inflammatory, antihypertensive, and cardioprotective activities	[39]
Synthesis apigenin	Mice 50 mg/kg•day by oral gavage for 4 weeks	Attenuated insulin resistance, dyslipidemia and liver injury, and mitigated oxidative stress	[40]
Pentacyclic triterpenes: oleanolic acid, 18β-glycyrrhetinic acid, ursolic acid, α, β-amyrin, carbenoxolone, asiatic acid, corosolic acid, bardoxolone methyl, lupeol, ilexgenin A	Rats and mice Oleanolic acid (25 mg/kg•day, once daily, 10 weeks) to fructose-fed rats; ursolic acid-treated fat-fed mice at a dose of 50 or 200 mg/kg of body weight (orally for 8 weeks); lupeol at 0.67 g/kg, given orally for 7 weeks	Decreased fatty acid synthesis, triglyceride synthesis and plasma triglycerides, leptin and free fatty acids, and also triglyceride content in skeletal muscle Reduced fatty liver, adipocyte size, hepatic steatosis, insulin resistance, inflammation, oxidative stress, body weight, atherosclerosis, and hypertension Decreased cholesterol synthesis, total cholesterol, VLDL and LDL-cholesterol, cholesterol in liver and in adipose tissue	[26]
Synthesis carvacrol	Mice Carvacrol was added to the diet in a 0.1% (w/w) (equivalent to 100 mg/kg body weight) for 10 weeks	Prevented obesity by decreasing body weight gain, visceral fat-pad weights, and plasma lipid levels; also inhibited visceral adipogenesis and attenuated the production of pro-inflammatory cytokines in visceral adipose tissues	[41]
Oryzanol and ferulic acid	Mice High-fat diet supplemented with 0.5% (w/w) oryzanol or 0.5% (w/w) ferulic acid for 7 weeks	Decreased in body weight, improved blood glucose metabolism, may be beneficial for the treatment of diabetic hyperglycemia	[42]

Phytochemical(s)	Methodology	Beneficial effect(s)	Reference
Curcuminoids	Rats Rats were fed with high-fat diets with curcuminoid supplement at concentrations of 30, 60, and 90 mg per kilogram of body weight every day for 12 weeks	Decreased plasma free fatty acid levels and improves cardiac autonomic nervous system activity in obesity Contributed to lower body fat and body weight gain Improved obesity-associated inflammation and associated metabolic disorders such as insulin resistance, hyperglycemia, hyperlipidemia, and hypercholesterolemia	[43]
Quercetin	Rats and mice Rats and mice fed with a Western diet containing 0.05% quercetin for 20 weeks	Decreased body weight, visceral fat, blood glucose, free cholesterol, total antioxidant status, lipid accumulation, and systolic blood pressure	[24]
Green tea catechins	Rats and mice Different dose-time treatment: from 0.5–4% of different catechins for 6–22 weeks	Decreased body weight, total lipids, cholesterol, and triglycerides in liver and plasma. Also improved glucose homeostasis: increased glucose tolerance and decreased serum glucose, insulin resistance, and homeostasis model assessment of insulin resistance	[27]

Table 2.
Fruits, vegetables (or extracts), and phytochemicals endorsed by recent in vivo (preclinical trials) studies.

Phytochemical(s)	Methodology	Beneficial effect(s)	Reference
Resveratrol	A randomized, double-blind, placebo-controlled clinical trial was carried out in 24 patients with diagnosis of metabolic syndrome 12 patients received trans-resveratrol (500 mg) three times per day before meals for 90 days	Decreased weight, body mass index, fat mass, waist circumference, and total insulin secretion	[44]
Artichoke leaf extracts rich in flavonoids and caffeoylquinic acid derivatives	Double-blind placebo-controlled randomized clinical trial was carried out in 80 patients with a diagnosis of metabolic syndrome 80 patients with metabolic syndrome received 1800 mg of artichoke leaf extract as four tablets per day for 12 weeks	Decreased ox-LDL and triglyceride levels	[45]
Oligonol extract	Randomized double-blind, placebo-controlled study with 18 subjects. All subjects took two capsules of Oligonol (50 mg/capsule) twice a day for 10 weeks.	Decreased body weight, abdominal circumference, and visceral fat volume	[46]
Pomegranate juice	A randomized, double-blind, placebo-controlled clinical trial was carried out in 20 obese [body mass index (BMI) 30.0–39.9] adults 10 patients received 120 mL of pomegranate juice or placebo while in a fasted state before breakfast every day for 1 month	Did not modify insulin secretion and sensitivity in patients with obesity; however, the natural evolution to increased weight and adiposity was halted	[47]

Phytochemical(s)	Methodology	Beneficial effect(s)	Reference
Cocoa extract	Double-blind, randomized, placebo-controlled parallel nutritional intervention with 50 obese volunteers [30.59(2.33) kg/m^2]. Meals supplemented with 1.4 g/day cocoa extract for 4 weeks	A marginal decrease ($P = 0.072$) in oxidized bases was observed, which attributed to weight loss	[48]
Green tea catechins	Various randomized, double-blind, placebo-controlled clinical trials. Some studies with different dose-time treatment: from 38 to 600 mg/day of different catechins for 6–24 weeks	Not all studies have found positive results for obesity-related measures. Green tea administration has also shown no influence on body weight, body mass index, fat mass, and waist and hip circumference	[27]
Black seeds and turmeric	Double-blind randomized controlled trial Black seeds (1.5 g/day), turmeric (2.4 g/day), its combination (900 mg black seeds and 1.5 g turmeric/day) for 8 weeks	Improved blood pressure, waist circumference, hip circumference, body mass index, LDL-cholesterol, HDL-cholesterol, and triglyceride content	[49]

Table 3.
Fruits, vegetables (or extracts), and phytochemicals endorsed by recent in vivo (clinical trials) studies.

In vitro studies achieve 50% of cell population growth inhibition with active compound amount of ppm, and, in some cases (some flavonoids, ajoene, catechins), a dose-time-dependent effect is detected. These dose-time-dependent effects, together with the multivariate effects observed in most cases, are largely due to the antioxidant properties of the phytochemicals considered. Flavonoids tested on cell population growth were well correlated with their antioxidant activity [36].

The doses used in the preclinical studies were much higher than those used in the clinical trials, and the limitations in the ingested doses may be the cause of the ineffective results. Kobori et al. [24] concluded that in vitro anti-aggregatory effects of flavonoids are caused by concentrations that cannot be attained in vivo by dietary consumption.

In most preclinical studies, imbalances are induced in mice with a high fructose diet, and the effect that the active component has on these induced disorders is determined. Also in the clinical studies performed with positive results, the component of interest was provided as part of a low-fat diet to patients with physiological alterations associated with metabolic syndrome. In these conditions the results show a palliative effect but in no case a preventive effect. Thus, quercetin supplementation did not affect the antioxidant status under healthy, normal conditions [24].

Sometimes solubility and bioavailability are a major limitation factor. As an example, pentacyclic triterpenes such as ursolic acid are poorly bioavailable because of poor aqueous solubility and permeability through biological membranes [26]. These limit their biological effects. Some strategies have increased bioavailability. For example, on comparing the oral bioavailability of ursolic acid microcrystals and nanocrystals to its coarse suspension in rats, ursolic acid microcrystals and nanocrystals exhibited 1.40- and 2.56-fold enhancement, respectively. Also, a new product from polyphenols of lychee has been developed. Oligonol, a unique low-molecular-weight polyphenol, was developed to enhance absorption of polyphenols from the intestines. It contains 15.7% polyphenol monomer ((+)-catechin, (−)-epicatechin, etc.) and 13.3% polyphenol dimer (procyanidin B2, etc.), while lychee fruit-derived polyphenol contains 6.4% polyphenol monomer and 9.9% polyphenol dimer [46].

Concluding from above, translational studies from animal observations to human clinical trials and ultimately community interventions are needed to further confirm the effects of phytochemicals and foods rich in these bioactive compounds. The technological development should be aimed at the implementation of strategies that increase the bioavailability of active components with proven activity and that allow to provide the adequate doses avoiding toxicity problems.

2.2 Probiotics

Probiotics are defined as live microorganisms which, when administered in adequate amounts, confer a health benefit on the host [50]. Evidence from the latest studies in which probiotic efficiency in metabolic disorder management is assessed by both in vitro and in animal or human subjects is compiled in **Tables 4** and 5.

Regarding the strains employed in the management of several inflammatory diseases, they usually belong to the *Lactobacillus* and *Bifidobacterium* genera, although *Pediococcus pentosaceus* LP28, *Bacteroides uniformis* CECT 7771, and *Akkermansia muciniphila* have also been proved to have anti-obesity effects [65]. In general, doses

Probiotic organism	Methodology	Beneficial effect(s)	Reference
Lactobacillus plantarum Ln4 isolated from napa cabbage kimchi	3T3-L1 adipocytes	Inhibited adipogenesis and stimulated glucose uptake	[51]
	Mice fed on a standard diet or a high-fat diet (5–7 mice per group) supplemented or not with 5×10^8 CFU/day for 5 weeks	Reduced diet-induced weight gain, lipid accumulation, and insulin resistance	
Lactobacillus reuteri 263 patented strain	10 rats fed on a normal diet and 30 rats fed on a high-energy diet supplemented or not with 2.1×10^9 CFU/rat/day or 1.05×10^{10} CFU/rat/day for 8 weeks	Reduced obesity by decreasing pro-inflammation factors and increasing antioxidant enzymes in the serum	[52]
Green tea (rich in epigallocatechin gallate) and *Houttuynia cordata* leaf (rich in chlorogenic acid) extract fermented with *Lactobacillus paracasei* subsp. *paracasei* NTU 101 originally isolated from infant	3T3-L1 pre-adipocyte model	Fermented tea powder promoted lipase activity in adipocytes, which thereby improves the lipolytic effect	[53]
	Rats fed for 8 weeks on a normal diet or a high-fat diet supplemented or not with: • Unfermented tea powder • (12.5 mg EGCG/day) Fermented tea powder • (12.5–25 mg EGCG/day plus 3.75–7 × 10^{10} CFU/day) NTU 101 powder • (7.5 × 10^{10} CFU/day) • EGCG powder (25 mg/day)	NTU 101-fermented tea had a more significant effect on the reduction of body weight gain and body fat content than the unfermented tea	

Probiotic organism	Methodology	Beneficial effect(s)	Reference
Lactobacillus fermentum strain 4B1 isolated from fermented rice and shrimp compared to a commonly prescribed weight loss drug	35 obese induced mice. Daily dose: none, 12 mg/kg orlistat (Xenical®) or 2.5 × 10^{10} CFU/kg for 21 days	Prevented obesity in lean hosts and reduced body weight gain and adipose tissue weight in mice receiving the high-fat diet	[54]
Heat-killed and live *Lactobacillus reuteri* GMNL-263	Rats fed for 12 weeks on a normal diet or a high-fat diet supplemented or not with 2 × 10^9 cells/day for 12 weeks	Both heat-killed and live cells prevented obesity, insulin resistance, and hepatosteatosis in high-fat diet rats by suppressing the inflammatory response and the expressions of specific cytokines	[55]
Live or pasteurized *Akkermansia muciniphila*	Obese and diabetic diet-induced mice	Pasteurization-enhanced bacterium capacity to reduce fat mass development, insulin resistance, and dyslipidemia induced by a high-fat diet	[56]
Bifidobacterium pseudocatenulatum SPM 1204, *Bifidobacterium longum* SPM 1205, and *Bifidobacterium longum* SPM 1207	36 rats fed for 5 weeks on a normal diet or a high-fat diet supplemented or not with 10^8–10^9 CFU(1:1:1)/day	Reduced body weight and fat gain, as well as total cholesterol, HDL-cholesterol, and LDL-cholesterol levels in serum blood and harmful enzyme activity	[57]
Lactobacillus reuteri ATCC PTA 4659, *Lactobacillus reuteri* DSM 17938, and *Lactobacillus reuteri* L6798	40 mice fed on a high-fat diet for 12 weeks supplemented or not with 10^9 CFU/day of a specific strain	Strain ATCC prevented obesity, lowered blood insulin level, and affected liver steatosis in hypercholesterolemic mice on a high-fat diet	[58]
Pediococcus pentosaceus LP28 isolated from longan fruit and *Lactobacillus plantarum* SN13T	5 lean control and 30 diet-induced obese mice fed for 6 weeks with a regular diet or a high-fat diet supplemented or not with 1.25 × 10^9 CFU/g of a specific strain	Live LP28 reduced body weight gain and liver lipid contents, whereas heat-killed and SN13T were ineffective	[59]

Table 4.
Probiotic strains endorsed by in vitro and animal studies.

greater than 10^8 CFU/day were orally administered to the drinking water or by oral gavage of the lyophilized bacterial powder in water (in animal studies) as well as in the form of capsules or fermented milk products (in human studies). Probiotic supplementation in rat and mouse studies usually applies to both diet-induced obese individuals and lean individuals fed on a high-fat diet, thus showing the effect of probiotics in both the treatment and the prevention of several metabolic abnormalities. However, human studies are basically applied to overweight or obese healthy individuals, submitted or not to energy restriction and/or regular exercise. As regards in vitro studies, treatment of 3T3-L1 pre-adipocytes with test substances during their differentiation stage is the most common technique.

Probiotic organism	Methodology	Beneficial effect(s)	Reference
DUOLAC 7 including *S. thermophilus* KCTC 11870BP, *L. plantarum* KCTC 10782BP, L. *acidophilus* KCTC 11906BP, *L. rhamnosus* KCTC 12202BP, *B. lactis* KCTC 11904BP *B. longum* KCTC 12200BP, and *B. breve* KCTC 12201BP (5×10^{12} CFU/capsule)	A randomized, double-blinded, placebo-controlled study in 50 female aged 19–65 with BMI > 25 kg/m^2 and waist circumference > 85 cm following usual dietary intake and lifestyle receiving Bofutsushosan (3 g per administration) and probiotics (1 capsule) or Bofutsushosan and placebo twice per day for 8 weeks	Probiotics increased HDL cholesterol level and effectively modified the composition of gut microbiota *Bifidobacterium breve* was the only strain showing a significant tendency of declination of endotoxin level, so it was suggested as a promising probiotic strain specified for obesity treatment	[60]
L. rhamnosus CGMCC1.3724 (LPR) in capsules (1.62×10^8 CFU/capsule) with a mix 70:30, v/v of oligofructose and inulin (300 mg/capsule)	A double-blind, placebo-controlled, randomized trial in 125 healthy overweight men and women (age between 18 and 55, BMI between 29 and 41 kg/m^2) following a supervised diet and consuming two capsules per day or placebo for 24 weeks	LPR supplementation accentuated body-weight loss in women submitted to energy restriction; this effect persisted in the subsequent maintenance phase, when energy restriction was not further imposed	[61]
L. plantarum TENSIA (DSM 21380) isolated from the gastrointestinal tract of healthy Estonian children added to cheese milk in amounts of 1.5×10^{11} CFU/g before renneting	A randomized, double-blind, placebo-controlled, parallel-designed study in 40 subjects with metabolic syndrome following a hypocaloric diet supplemented with 50 g/day of probiotic or control cheese for 3 weeks	The hypocaloric diet supplemented with the probiotic cheese reduced BMI, arterial blood pressure, and the risk of metabolic syndrome in obese patients with hypertension	[62]
Milk fermented with or without *Lactobacillus gasseri* SBT2055 (LG2055)	Multicenter, double-blind, placebo-controlled intervention trial in which 87 subjects (BMI of 24.2–30.7 kg/m^2 and abdominal visceral fat area of 81.2–178.5 cm^2) were randomly assigned to consume 200 g/day of fermented milk with or without LG2055 for 12 weeks while maintaining their habitual mode of living	Intake of the probiotic LG2055 reduced abdominal visceral and subcutaneous fat areas as well as body weight, BMI, waist and hip circumferences, and body fat mass	[63]
Lactobacillus amylovorus and *Lactobacillus fermentum* microencapsulated in yogurt (1.39×10^9 CFU/ yogurt)	A placebo-controlled, double-blind crossover clinical investigation with 28 healthy but overweight individuals (BMI between 25 and 32 kg/m^2)	Probiotic consumption altered intestinal microflora in a manner that was associated with reduced total body adiposity, an important anthropometric indicator of obesity	[64]

Table 5.
Probiotic strains endorsed by in vivo studies.

Among the analytical determinations, the most representative are changes in the body weight and the body fat content as well as main serum biochemical parameters (glucose, insulin, leptin, lipids, lipoproteins, and inflammatory indicators). Postmortem determinations, such as the liver weight or the adipose tissue histology, are also common in animal studies. Finally, since probiotics are known to increase the bacterial diversity of intestinal microflora, in vivo studies usually include viable counts in fecal samples and evaluation of intestinal survival. In fact, a lot of recent research relates gut microbiota composition with almost every chronic disease: from gastrointestinal diseases to obesity, diabetes, cancer, and even neurological and neurodegenerative disorders such as depression, autism, anxiety, and Parkinson's disease [66]. Not only does a certain microbiota predispose to suffer certain diseases, but also the incidence of a certain disorder modifies the gut microbiota of an individual. In overweight/obese subjects, *Bacteroides*, *Parabacteroides*, *Ruminococcus*, *Campylobacter*, *Dialister*, *Porphyromonas*, *Staphylococcus*, and *Anaerostipes* are the dominant genera linked to a low diversity of species, while *Faecalibacterium*, *Bifidobacterium*, *Lactobacillus*, *Butyrivibrio*, *Alistipes*, *Akkermansia*, *Coprococcus*, and *Methanobrevibacter* are predominant in lean individuals with a high bacterial diversity [67]. Apparently, the intestinal microflora of obese subjects is more efficient at extracting energy from a given diet than that of lean individuals, thus leading to increased energy storage and adiposity [65]. Moreover, beneficial intestinal microflora is known to produce short-chain fatty acids (e.g., acetate, butyrate, and propionate) from indigestible polysaccharides, which may act as energy substrates as well as regulators of satiety and food intake. Last but not least, *Lactobacilli* and *Bifidobacteria* are known to synthesize bioactive isomers of conjugated linoleic acid with antidiabetic, anti-atherosclerotic, immunomodulatory, and anti-obesity properties [68]. In other words, low bacterial diversity in obese individuals is associated with a reduction in butyrate-producing bacteria, a reduction in hydrogen and methane production, an increase in mucus degradation, and an increase in the potential to manage oxidative stress. Since intestinal microflora composition is strongly affected by dietary patterns, studies evaluating the effect of certain food components on the growth of bacteria with beneficial effect on metabolic syndrome and obesity, particularly *Akkermansia muciniphila* and *Faecalibacterium prausnitzii*, are of great interest. Increasing the intestinal population of these two species has become a real opportunity to decrease alterations associated with obesity and metabolic disorders [69]. In addition to this, high-fat diet treatment has been proven to induce metabolic changes that impair gut barrier function in rats [55].

Together with increasing gut microbiota diversity, the production by fermentative action of those bioactive molecules involved in the metabolic pathways that trigger the metabolic syndrome has also taken a lot of interest in the last years. In particular, many studies focus on the phenolic compound bioconversion by food fermentation into other components with greater beneficial effect on the abnormalities associated with metabolic syndrome. As an example, Wang et al. [53] proved that the use of *Lactobacillus paracasei* subsp. *paracasei* strain NTU 101 in the fermentation of green tea and *Houttuynia cordata* leaves increased the levels of epigallocatechin gallate (EGCG), epigallocatechin (EGC), and chlorogenic acid, which enhanced the probiotic effect on body fat reduction. These results show the synergistic or complementary effect between the two bioactive compounds: the probiotic strain increases gut microbiota diversity and enhances intestinal absorption, while the EGCG acid promotes the lipolysis process. Zarrati et al. [70] also reported a synergistic effect between a weight loss diet and probiotic yogurt in overweight and obese individuals.

It should be noted that in order to exert their health benefits, probiotics do not necessarily have to be alive. In fact, heat-killed *Lactobacillus reuteri* GMNL 263 was as effective as live *Lactobacillus reuteri* GMNL 263 in attenuating obesity-induced metabolic abnormalities in high-fat diet-induced rats by reducing insulin resistance and hepatic steatosis formation [55]. Also both heat-killed *Lactobacillus plantarum* strain Ln4 and freeze-dried cultured MRS broth significantly reduced lipid accumulation and stimulated glucose uptake in 3T3-L1 adipocytes [51]. Finally, Plovier et al. [56] found that pasteurization enhanced the capacity of *Akkermansia muciniphila* to reduce fat mass development, insulin resistance, and dyslipidemia in mice. It seems that a specific protein isolated from the outer membrane of *Akkermansia muciniphila* is stable at temperatures used for pasteurization and improves the gut barrier, thus being the main responsible factor of the beneficial effect of the bacteria on health.

Of all the studies analyzed, it is concluded that many microorganisms have the potential for development as therapeutic probiotics for obesity and associated disorders. However, due to the strain specificity of probiotic microbes in exerting their beneficial effects, bacterial strains of the same species have different effects on adiposity and insulin sensitivity.

3. Case study: functional food development from lulo fruit with potential effect against metabolic syndrome

The lulo fruit (*Solanum quitoense* Lam.), also known as "naranjilla," is an important native Andean crop. Grown and consumed mainly in Colombia, Ecuador, and Central America, the plant produces a spherical, 3–8-cm-diameter fruit with orange skin (epicarp) covered by short, stiff, and thorny hairs or spines. The internal structure of the fruit is similar to that of the tomato fruit: a very juicy, acidic, and translucent yellow-green pulp (mesocarp and endocarp) that is located in four compartments separated by membranous partitions [71]. In Colombia, lulo is an economically important crop which, in 2015, was grown in a total area of 10,623 ha, with a total yield of 82,354 tons and an average yield of 9.6 tons/ha [72]. Although the principal market of this crop is in the producing countries themselves, it has gained interest in recent years in national and international markets due to its organoleptic properties and its nutritional value. In fact, lulo has an intense and refreshing taste and is rich in proteins, vitamin C, fiber, and antioxidant compounds, such as all-trans-β-carotene, lutein and zeaxanthin, chlorogenic acids, and flavonol glycosides [73–76], in addition to iron, calcium, phosphorus, and some precursors of vitamin A [77] (**Table 6** [78]). In particular, fruit carotenoids present in lulo fruit have been associated to the prevention of several illnesses, including hypertension, obesity, and cardiovascular diseases [79–81]. Also the potential of lulo as an antihypertensive agent is related to its content in N1,N4,N8-tris (dihydrocaffeoyl) spermidine and N1,N8-bis (dihydrocaffeoyl) spermidine (actually bioactive amines), which are bitter active compounds with inhibitory activity against the angiotensin-converting enzyme (ACE-1) that indirectly increases the blood pressure by causing blood vessels to constrict [35]. In turn, when evaluating the antihypertensive activity of some compounds of the lulo fruit by means of chemical computation techniques, the researchers of the Natural Additives of Aroma and Color group from the Chemistry Department of the Universidad Nacional de Colombia found that this was between 10 and 20 times higher than that of the drugs traditionally used to treat hypertension.

Based on the considerations made above, the lulo fruit comes to be a promising alternative with regard to the prevention and relief of hypertension-related diseases. However, being a highly perishable fruit, technological transformation processes are indispensable to take advantage of its beneficial properties by as many consumers as possible. According to this, the group of Functional Foods of the University Institute of Food Engineering for Development of the Polytechnic

Proximates		Minerals		Vitamins	
Carb	5.9 g	Ca	8 mg	Folate	3 µg
Fiber	1.1 g	Mg	11 mg	Vit B3	1.45 mg
Protein	0.44 g	P	12 mg	Vit B5	0.22 mg
Sugars	3.74 g	K	200 mg	Vit B1	0.05 mg
Fat	0.22 g	Na	4 mg	Vit B6	0.11 mg
Water	93.05 g	Cu	0.03 mg	Vit C	3.2 mg
Energy	25 cal	Fe	0.35 mg	α-Carotene	4 µm
		Mn	0.07 mg	β-Carotene	333 µm
		Se	0.4 µg	β-Cryptoxanthin	10 µm
		Zn	0.1 mg	Lutein and zeaxanthin	299 µm
				γ-Tocopherol	0.2 mg
				α-Tocotrienol	0.01 mg
				γ-Tocotrienol	0.01 mg
				Vit A	28 µm
				Vit E	0.75 mg
				Vit K	14.6 µ

Values expressed for 100 g of fresh fruit.

Table 6.
Nutrition facts of lulo fruit.

University of Valencia (Spain), in conjunction with the group of Biodiversity Evaluation and Use of the Technological University of Chocó (Colombia), is working on the development of new functional foods from lulo fruit (*Solanum quitoense* Lam). On the one hand, a stable lulo juice with improved antioxidant properties has been obtained by means of the application of moderate high homogenization pressures (from 50 to 150 MPa) instead of traditional thermal pasteurization. The same juice has proved to be a suitable impregnation liquid for the enrichment of other fruits with a porous structure, such as Granny Smith apples. The lulo fruit itself was found to have a high impregnation capacity, which implies susceptibility to be enriched with different bioactive compounds. Finally, after fermentation with *Lactobacillus reuteri*, selected for being one of the strains with proven effect against the metabolic syndrome, the number of viable counts in lulo juice resulted to be high enough to claim that it also may exert a probiotic effect. Most relevant results in relation to these advances are shown next.

3.1 Enhancing antioxidant properties of lulo juice by means of moderate high-pressure homogenization

This section shows the effect that homogenization pressures in the range of 50–150 MPa have on main physicochemical properties, including the total content of phenols and flavonoids and the antioxidant activity measured by both DPPH and ABTS methods. To obtain the juice, washed and without peduncle lulo fruits were crushed for 10 min in a blender (Phillips Avance Collection Standmixer, 800W 2L). The liquefied product was then filtered with a stainless steel sieve of 500 µm nominal aperture. When necessary, the juice was homogenized at 50, 100 or 150 MPa in a laboratory scale high-pressure homogenizer (Panda Plus 2000, GEA-Niro Soavi, Parma, Italy).

Homogenization pressure	Brix	pH	ρ (g/cm³)	K (Pa sn)	n
0 MPa	6.57 ± 0.12^a	3.13 ± 0.02^a	1.04 ± 0.02^a	0.39 ± 0.12^a	0.44 ± 0.06^b
50 MPa	6.4 ± 0.4^a	3.12 ± 0.02^a	1.06 ± 0.04^a	0.9 ± 0.4^b	0.37 ± 0.04^a
100 MPa	6.33 ± 0.15^a	3.18 ± 0.03^a	1.07 ± 0.02^a	0.79 ± 0.02^{ab}	0.37 ± 0.03^a
150 MPa	6.4 ± 0.4^a	3.15 ± 0.02^a	1.09 ± 0.02^a	1.3 ± 0.5^b	0.34 ± 0.04^a

[abc...] *different superscripts in the same column indicate statistically significant differences ($p < 0.05$).*

Table 7.
Effect of homogenization pressure on pH, soluble solid content (brix), apparent density (ρ), and rheological properties of lulo juice.

As it can be observed in **Table 7**, neither the soluble solid content nor the pH or the density of the lulo juices was significantly affected by homogenization pressure. On the contrary, the consistency index (K) increased significantly after the homogenization step, which is directly related to particle size reduction. As regards the average size of particles, it was maximum in the non-homogenized juice (251 ± 5 μm) and minimum in the juice homogenized at 150 MPa (57.94 ± 0.14 μm). Therefore, homogenization increased the amount of solids in suspension and, consequently, the stability of the cloud.

As regards the antioxidant properties of lulo juice, the fruit's own transformation into juice significantly reduced both total phenol and total flavonoid contents, which were probably separated from the juice together with the bagasse during the filtration step. However, the concentration of such compounds increased slightly (from 1.03 ± 0.16 to 1.28 ± 0.07 mg GAE/g for phenols and from 0.35 ± 0.24 to 0.570 ± 0.011 mg QE/g for flavonoids) after juice homogenization at 150 MPa and the subsequent reduction in the average particle size. Similar trends were observed when analyzing the total antioxidant activity by both the ABTS and the DPPH methods and when quantifying spermidine by HPLC analysis. For the latter compound, concentration increased from 1.86 ppm in non-homogenized lulo juice to 2.04 ppm in lulo juice homogenized at 150 MPa.

Regarding the ability of the homogenized lulo juice treated at 150 MPa to impregnate Granny Smith apples sliced, it was found to be similar to that of an isotonic sucrose solution. In this way, the bioactive compounds present in lulo juice may become part of the vacuum impregnated fruit composition. To be more precise, around 0.22 m^3 of lulo juice homogenized at 150 MPa could be incorporated to every m^3 of fresh apple.

3.2 Lulo fruit as a food matrix for vacuum impregnation and food property improvement

In this section, impregnation properties of lulo fruit are discussed. Vacuum impregnation being a matrix engineering technique allows to introduce desirable compounds into the porous structure of foods by applying a pressure gradient [14]. Among the impregnation parameters, the volume of the external liquid that can be incorporated into the cellular tissue in a controlled way (i.e., X, in m^3/m^3) stands out, which informs about the feasibility of incorporating physiologically active compounds into its porous structure for the formulation of new products with enhanced functional properties. Hence, unpeeled lulo fruit was cut into 5 mm thick slices and immersed in an isotonic sucrose solution ($a_w = 0.994 \pm 0.003$). Vacuum impregnation was carried out in a pilot plant scale equipment located at the University Institute of Food Engineering for Development of the Universitat Politècnica de

València (Spain). This equipment consists of a stainless steel vacuum chamber connected to a liquid ring pump (SIHI model LOHE-25007). The vessel containing the impregnating solution was placed into the vacuum chamber, and the lulo samples were immersed in the liquid by means of a pneumatic arm operated by a compressor (COMBA, 1,5 HP de 25 L). The working conditions were set at 50 mbar for 10 min and atmospheric pressure for 10 min more. In each trial, the weight change of the samples was recorded according to the procedure described by [82], thus allowing to calculate the characteristic impregnation parameters of the lulo fruit.

As it is shown in **Table 8**, the different batches analyzed behaved in a similar way during the vacuum impregnation step. Positive values of parameters X_1 (between 1 and 5%) and X (between 8.6 and 16%) indicate that the impregnating liquid entered the porous structure after both the vacuum and the atmospheric steps. Likewise, positive values of parameters γ_1 (between 3.9 and 7.1%) and γ (between 2.9 and 6.6%) indicate a volumetric expansion of the lulo matrix after both the vacuum and the atmospheric steps. Compared to other fruits and vegetables [83], the volume fraction of fresh lulo that was filled with the impregnating solution at the end of the process (X, in $m^3/100\ m^3$) was significantly lower than that of Granny Smith apple (21.0 ± 0.9) or Soraya aubergine (64 ± 2) but considerably higher than that of Chandler strawberry (6.4 ± 0.3), Hayward kiwifruit (0.7 ± 0.5), or Bulida apricot (2.2 ± 0.2). Despite such differences, the lulo matrix can be considered as suitable to be enriched with other active compounds by means of the vacuum impregnation technique.

3.3 Probiotic food development from lulo fruit

The growing number of consumers with lactose intolerance, high cholesterol levels, and/or following vegetarian or vegan diets has encouraged the recent use of fruits and vegetables as probiotic carriers in the development of new functional foods. Fruit and vegetable juices are especially suitable for the growing of probiotic microorganisms since they inherently contain beneficial nutrients and have taste profiles that are pleasing to all the age groups [84]. In addition, due to their fast passage through the digestive tract, the viability of probiotic cells in the juices is hardly affected by the harsh acidic environment of stomach [85]. However, these food matrices do not always fulfill the pH or the essential amino acids and vitamins required for the optimum growth of most LAB with proven probiotic effect. This section evaluates the possibility of using the non-homogenized lulo juice as a medium for the growth of *Lactobacillus reuteri* CECT 925T. For this purpose, the lulo juice obtained by the procedure described above was pasteurized at 75°C for

Batch	X_1	γ_1	X	γ	ε_e
1	5 ± 7^a	5 ± 4^a	8.8 ± 1.6^a	3 ± 3^a	6 ± 4^a
2	2.1 ± 1.8^a	6 ± 5^a	10 ± 3^a	5 ± 4^a	6 ± 5^a
3	2 ± 4^a	5 ± 2^a	11 ± 2^{ab}	3.7 ± 0.9^a	8 ± 2^a
4	1 ± 1.4^a	3.9 ± 0.9^a	16 ± 6^b	6.6 ± 1.0^a	9 ± 6^a
5	2.5 ± 1.3^a	7.1 ± 1.0^a	8.6 ± 0.9^a	2.9 ± 0.8^a	6.3 ± 1.2^a

[abc...] different superscripts in the same column indicate statistically significant differences ($p \leq 0.05$).
X_1 and X stand for the volume fraction of fresh sample impregnated at the end of the vacuum step and at the end of the atmospheric step, respectively; γ_1 and γ stand for the relative volume deformation of fresh sample at the end of the vacuum step and at the end of the atmospheric step, respectively; ε_e stands for the effective porosity.

Table 8.
Vacuum impregnation response of lulo fruit slices (5 mm thick).

2.5 min before being inoculated with 4 mL/L of MRS broth containing the active microorganism in a concentration of 10^8 CFU/mL. After 24 hours of incubation at 37°C, viable counts in the juice were of the order of 10^6 CFU/mL. Although this value was high enough to make an EU-based health claim [86], it was significantly lower to that obtained in mandarin juice inoculated with either *Lactobacillus salivarius* spp. *salivarius* CECT 4063 or *Lactobacillus acidophilus* CECT 903 [87].

In a further step, the lulo juice containing the probiotic was employed as impregnating liquid for the vacuum impregnation of Granny Smith apple slices (5 mm thick). In this way, the probiotic was introduced into a solid matrix without disturbing its organized cellular structure. However, since only 20% of the initial volume of the apple is filled with the impregnation liquid during the vacuum impregnation step, the probiotic content in the impregnated apple was not greater than 10^5 CFU/g. Subsequent lyophilisation of the vacuum impregnated apples did not increase the *Lactobacillus reuteri* content as expected by water removal and subsequent weight loss, it being lower than 10^6 CFU/g in the liophylized sample. Probiotic counts in both the lulo juice and the impregnated apple snack could be improved by adding certain ingredients (e.g., prebiotics, cryprotectants, soygerm powder, yeast extract, etc.) and/or applying specific processing technologies that can improve microorganism survival such as microencapsulation or sublethal homogenization. In any case, it should be interesting to evaluate through both in vivo and in vitro studies the antihypertensive activity of *Lactobacillus reuteri* in the products designed, since it could be enhanced due to a synergistic effect with the spermidine from the lulo juice.

4. Market and consumer trends toward functional foods

Revolution in living standard, eating habits, and increased health awareness has shifted consumer's acceptance toward nutritious, healthy, and disease-preventive food with wider health benefits. Consumer is becoming more and more conscious about the role of food in life extension, well-being, and prevention of chronic diseases [87].

Specific consumer characteristics, such as demographic background or personal motivation to participate in pro-health activities, play a remarkable role in functional food acceptance and consumption. Some sociodemographic characteristics such as gender, education, and age are the most important factors related to the acceptance of functional food. In addition, apart from health benefits, the carrier and the origin of functional components play an important role in making the decision to purchase functional products, consumers being more likely to purchase those functional components found naturally in foods. Other factors, such as organoleptic attributes, convenience, or label information, are found to be essential for consumer's acceptance. In his study, Kraus [88] concludes that consumers are not willing to sacrifice taste and general pleasure of eating and also states that naturalness of a product is very important.

Particularly for probiotics, a major challenge for these products is product acceptability by consumers with regard to sensory criteria. Traditionally, health benefits of probiotics were based in the consumption of fermented dairy products; however, lactose intolerance, cholesterol content, and allergic milk proteins have limited the growth of dairy probiotics. Besides, the increase in vegetarian consumers in both developed and developing countries has also contributed to a growing demand for plant-based probiotic products [87]. According to Panghal et al. [89], fruits are healthy and refreshing and have good taste and flavor profile and can be suitable for probiotics. They are an ideal medium to develop functional foods and have more nutritional values due to the presence of various phytochemicals, antioxidants, no cholesterol, vitamins, mineral content, and dietary fibers. Besides, economic reasons for the developing countries also require the search for an alternative to dairy products with good nutrients along with health-promoting factors, e.g., fruits, vegetables, cereal, legume, etc., and products which lack cholesterol content but are rich in protein, starches, minerals, fiber, vitamins, and antioxidants.

Nowadays an increasing trend in the Western society is consumer interest and focus toward natural and organic products, where the use of synthetic additives is limited. It has been suggested that natural ingredients with strong antioxidant activity could be used to design novel functional beverages. An increased interest relies upon the fortification with polyphenols due to their beneficial role against cardiovascular diseases, type 2 diabetes, and obesity, among other conditions. The combination of prebiotics, and also phenols with probiotic microorganisms, represents an innovative biotechnology to enlarge the functional food market and especially beverages [90].

According to Grand View Research [91], the global functional food market was higher than 129 billion dollars (US) in 2015, and it is expected to increase up to 250 billion in 2024. Growing consciousness among consumers on their health and proper diet, together with the prospect of reducing or even eliminating nutrition-related diseases, is responsible of this market trend. Society is becoming more and more conscious on the impact that changing dietary patterns may have in the incidence of type 2 diabetes, coronary heart disease, cancer, periodontal disease, and obesity. In this regard, functional foods are believed to play an outstanding role. In addition, increasing healthcare cost, along with the desire of improving later years among the geriatric population, has driven the growth of the functional food industry worldwide.

The global functional food market includes that of carotenoids, dietary fibers, fatty acids, minerals, prebiotics and probiotics, vitamins, minerals, phytochemicals, enzymes, and antioxidants in general. Market revenue of all these products separately is also expected to increase in the coming years. For example, dietary fibers, which are considered to prevent obesity and diabetes, are expected to grow by 8.4% in the next 8 years. Other phytochemicals, such as flavonoids, held a share of over 30% in terms of market value. Although these have been commonly consumed in their natural form, consumer's habits have led to their use in the form of functional food products which are aimed at preventing diet-related chronic diseases including those related to the metabolic syndrome. North America accounts for the largest market in flavonoids, the Asia Pacific demand was over 110 million US Dollars in 2015, and Europe is expected to grow, although at a slower pace. In any case, prevalence of diabetes, obesity, and chronic diseases is likely to propel demand for these nutritional foods and beverages in Europe.

With regard to probiotics, there is also a growing concern on awareness in their functional health benefits against different conditions, including those related to the metabolic syndrome such as obesity or type 2 diabetes [92]. The global probiotic market was thought to be worth 35.5 billion dollars in 2016, with predictions of this increasing up to 65 billion dollars by 2024 [93]. As reported by Lumina intelligence [92], a survey of Ganeden on consumers concluded that almost 80% of consumers preferred to consume probiotics in food and beverage products than in supplements. This is of special value taking into account that consumer preference is a key currency for measuring product success and predicting upcoming tendencies. North America demand for probiotics is expected to increase by 7.9% from 2016 to 2025, whereas the European market will grow at a pace of 7.3%. As for Asia Pacific countries, the probiotics industry is also expected to increase significantly.

Probiotics have achieved a prominent position in the global food market. Among the countries that have shown growth in the probiotic market, Europe represents the largest and fastest growing market, followed by Japan. Currently, there is a wide range of probiotic products offered by companies such as BioGaia Biologics AB, Christian Hansen A/S, ConAgra Functional Foods, Danisco, Groupe Danone, or Lifeway [87].

5. Conclusions

The huge increase in obesity and consequently of physiological disorders associated with this has led to a massive increase in research work conducted in this area over the past 10 years. The relationship between diet and the incidence of metabolic syndrome is clearly contrasted. Although this relationship is tremendously complex and it is hardly affected by other variables related to lifestyle, specific works establish phytochemicals and probiotics as two of the active components present in food, which have the greatest effect on prevention and in the reduction of symptoms associated with metabolic syndrome.

Currently, the technological development achieved by the food industry allows both the design and development of specific foods that include active components in their composition as well as the application of specific techniques that increase the functional value of natural foods. The use of these advances in the right direction can be decisive in the solution of health problems related to obesity. Specifically, the applications of moderate homogenization pressures or food formulation techniques, such as vacuum impregnation, are presented as possibilities to develop liquid and/or solid foods that combine the presence of phytochemicals and probiotics with demonstrated effectiveness against obesity in natural foods such as lulo fruit.

Acknowledgements

This research was funded in part by a PhD Research Fellowship from the Technological University of Chocó (Colombia). Authors specially thank Alicia Ríos Hurtado for her support, to the Government of the Department of Chocó and to the Research Group on Assessment and Use of Biodiversity (Chocó, Colombia).

Conflict of interest

Authors of this book chapter state that they do not have conflict of interest to declare.

Author details

Noelia Betoret[1], Leidy Indira Hinestroza[2], Lucía Seguí[1] and Cristina Barrera[1*]

1 University Institute of Food Engineering for Development, Universitat Politècnica de València, Valencia, Spain

2 Research Group on Assessment and Use of Biodiversity, Technological University of Chocó, Colombia

*Address all correspondence to: mcbarpu@tal.upv.es

References

[1] World Health Organization. 10 Facts on Obesity [Internet]. 2017. Available from: https://www.who.int/features/factfiles/obesity/en/ [Accessed: December 15, 2018]

[2] Park SY, Seong KS, Lim SD. Anti-obesity effect of yogurt fermented by *Lactobacillus plantarum* Q180 in diet-induced obese rats. Korean Journal for Food Science of Animal Resources. 2016;**36**(1):77-83. DOI: 10.5851/kosfa.2016.36.1.77

[3] Kaur J. A comprehensive review on metabolic syndrome. Cardiology Research and Practice. 2014;**2014**:1-21. DOI: 10.1155/2014/943162

[4] Esser N, Legrand-Poels S, Piette J, Scheen AJ, Paquot N. Inflammation as a link between obesity, metabolic syndrome and type 2 diabetes. Diabetes Research and Clinical Practice. 2014;**105**(2):141-150. DOI: 10.1016/j.diabres.2014.04.006

[5] Monteiro R, Azevedo I. Chronic inflammation in obesity and the metabolic syndrome. Mediators of Inflammation. 2010;**2010**:1-10. DOI: 10.1155/2010/289645

[6] Faloia E, Michetti G, De Robertis M, Luconi MP, Furlani G, Boscaro M. Inflammation as a link between obesity and metabolic syndrome. Journal of Nutrition and Metabolism. 2012;**2012**:1-7. DOI: 10.1155/2012/476380

[7] Liang Y, Hou Y, Niu H, Lu M, Xue L, Sun Q. Correlation of high-sensitivity C-reactive protein and carotid plaques with coronary artery disease in elderly patients. Experimental and Therapeutic Medicine. 2015;**10**(1):275-278. DOI: 10.3892/etm.2015.2486

[8] Flannagan KS, Ramírez-Zea M, Roman AV, Das AK, Villamor E. Adipose tissue polyunsaturated fatty acids and metabolic syndrome among adult parents and their children. Nutrition, Metabolism, and Cardiovascular Diseases. 2018;**28**(12):1237-1244. DOI: 10.1016/j.numecd.2018.08.008

[9] Retondario A, Fernandes R, Rockenbach G, Alves MA, Bricarello LP, Trindade EBSM, Vasconcelos FAG. Selenium intake and metabolic syndrome: A systematic review. Clinical Nutrition. 2019;**38**(2):603-614. DOI: 10.1016/j.clnu.2018.02.021

[10] Guarnizo-Poma M, Urrunaga-Pastor D, Montero-Suyo C, Lazaro-Alcantara H, Paico-Palacios S, Pantoja-Torres B, et al. Association between serum vitamin B12 levels and metabolic syndrome in a euthyroid population. Diabetes and Metabolic Syndrome: Clinical Research and Reviews. 2018;**12**(6):943-948. DOI: 10.1016/j.dsx.2018.05.022

[11] Urrunaga-Pastor D, Guarnizo-Poma M, Macollunco-Flores P, Lazaro-Alcantara H, Paico-Palacios S, Pantoja-Torres B, et al. Association between vitamin D deficiency and insulin resistance markers in euthyroid non-diabetic individuals. Diabetes and Metabolic Syndrome: Clinical Research and Reviews. 2019;**13**(1):258-263. DOI: 10.1016/j.dsx.2018.09.008

[12] Ford ES, Mokdad AH, Giles WH, Brown DW. The metabolic syndrome and antioxidant concentrations findings from the third National Health and nutrition examination survey. Diabetes. 2003;**52**(9):2346-2352

[13] Cardona-Velásquez S, Guzmán-Vivares L, Cardona-Arias JA. Systematization of clinical trials related to treatment of metabolic syndrome, 1980-2015. Endocrinología, Diabetes y Nutrición. 2017;**64**(2):82-91. DOI: 10.1016/j.endien.2016.09.004

[14] Betoret E, Betoret N, Vidal D, Fito P. Functional foods development: Trends and technologies. Trends in Food Science & Technology. 2011;**22**(9):498-508. DOI: 10.1016/j.tifs.2011.05.004

[15] Chen JP, Chen GC, Wang XP, Qin L, Bai Y. Dietary fiber and metabolic syndrome: A meta-analysis and review of related mechanisms. Nutrients. 2018;**10**(1):17. DOI: 10.3390/nu10010024

[16] Morrison DJ, Preston T. Formation of short chain fatty acids by the gut microbiota and their impact on human metabolism. Gut Microbes. 2016;**7**(3):189-200. DOI: 10.1080/19490976.2015.1134082

[17] Tassinari S, Azuero A, Arreaza D, Rueda-Rodríguez MC, Castañeda-Cardona C, Rosselli D. Suplementos nutricionales como modificadores del riesgo cardiovascular en componentes del síndrome metabólico en adultos. Revista Colombiana de Cardiología. 2017;**24**(3):277-285. DOI: 10.1016/j.rccar.2016.09.013

[18] Traber MG, Buettner GR, Bruno RS. The relationship between vitamin C status, the GUT-liver axis, and metabolic syndrome. Redox Biology. 2019;**21**:1-10. DOI: 10.1016/j.redox.2018.101091

[19] Iwaniak A, Darewicz M, Minkiewicz P. Peptides derived from foods as supportive diet components in the prevention of the metabolic syndrome. Comprehensive Reviews in Food Science and Food Safety. 2018;**14**:63-81. DOI: 10.1111/1541-4337.12321

[20] Khan MI, Anjum FM, Sohaib M, Sameen A. Tackling metabolic syndrome by functional foods. Reviews in Endocrine & Metabolic Disorders. 2013;**14**(3):287-297. DOI: 10.1007/s11154-013-9270-8

[21] Zemel MB. Role of calcium and dairy products in energy partitioning and weight management. The American Journal of Clinical Nutrition. 2004;**79**(5):907S-912S. DOI: 10.1093/ajcn/79.5.907S

[22] Shi H, Halvorsen YD, Ellis PN, Wilkinson WO, Zemel MB. Role of intracellular calcium in human adipocyte differentiation. Physiological Genomics. 2000;**3**(2):75-82. DOI: 10.1152/physiolgenomics.2000.3.2.75

[23] Zheng H, Lenard NR, Shin AC, Berthoud HR. Appetite control and energy balance regulation in the modern world: Reward-driven brain overrides repletion signals. International Journal of Obesity. 2009;**33**(S2):S8-S13. DOI: 10.1038/ijo.2009.65

[24] Kobori M. Dietary quercetin and other polyphenols: Attenuation of obesity. In: Watson RR, Preedy VR, Zibadi S, editors. Polyphenols in Human Health and Disease. Vol. 1. Oxford, UK: Academic Press; 2014. pp. 163-175. DOI: 10.1016/B978-0-12-398456-2.00014-1

[25] Pastor-Villaescusa B, Sánchez-Rodriguez E, Rangel-Huerta OD. Polyphenols in obesity and metabolic syndrome. In: Martí del Moral A, Aguilera-García CM, editors. Obesity: Oxidative Stress and Dietary Antioxidants. Oxford, UK: Academic Press; 2018. pp. 213-239. DOI: 10.1016/B978-0-12-812504-5.00011-8

[26] Sharma H, Kumar P, Deshmukh RR, Bishayee A, Kumar S. Pentacyclic triterpenes: New tools to fight metabolic syndrome. Phytomedicine. 2018;**50**:166-177. DOI: 10.1016/j.phymed.2018.09.011

[27] Wang S, Moustaid-Moussa N, Chen L, Mo H, Shastri A, Su R, et al. Novel insights of dietary polyphenols and obesity. Journal of Nutritional Biochemistry. 2014;**25**(1):1-18. DOI: 10.1016/j.jnutbio.2013.09.001

[28] Williams DJ, Edwards D, Hamernig I, Jian L, James AP, Johnson SK, et al.

Vegetables containing phytochemicals with potential anti-obesity properties: A review. Food Research International. 2013;**52**(1):323-333. DOI: 10.1016/j.foodres.2013.03.015

[29] Mallappa RH, Rokana N, Duary RK, Panwar H, Batish VK, Grover S. Management of metabolic syndrome through probiotic and prebiotic interventions. Indian Journal of Endocrinology and Metabolism. 2012;**16**(1):20-27. DOI: 10.4103/2230-8210.91178

[30] Arts IC, Hollman PC. Polyphenols and disease risk in epidemiologic studies. The American Journal of Clinical Nutrition. 2005;**81**(1):317S-325S. DOI: 10.1093/ajcn/81.1.317S

[31] Holubková A, Penesová A, Šturdík E, Mošovská S, Mikušová L. Phytochemicals with potential effects in metabolic syndrome prevention and therapy. Acta Chimica Slovaca. 2012;**5**(2):186-199. DOI: 10.2478/v10188-012-0029-8

[32] Richard AJ, Amini-Vaughan Z, Ribnicky DM, Stephens JM. Naringenin inhibits adipogenesis and reduces insulin sensitivity and adiponectin expression in adipocytes. Evidence-based Complementary and Alternative Medicine. 2013;**2013**:1-10. DOI: 10.1155/2013/549750

[33] Liang J, Halipu Y, Hu F, Yakeya B, Chen W, Zhang H, et al. Naringenin protects keratinocytes from oxidative stress injury via inhibition of the NOD2-mediated NF-κB pathway in pemphigus vulgaris. Biomedicine and Pharmacotherapy. 2017;**92**:796-801. DOI: 10.1016/j.biopha.2017.05.112

[34] Landa P, Kokoska L, Pribylova M, Vanek T, Marsik P. In vitro anti-inflammatory activity of carvacrol: Inhibitory effect on COX-2 catalyzed prostaglandin E2 biosynthesis. Archives of Pharmacal Research. 2009;**32**(1):75-78. DOI: 10.1007/s12272-009-1120-6

[35] Forero DP, Masatani C, Fujimoto Y, Coy-Barrera E, Peterson D, Osorio C. Spermidine derivatives in Lulo (*Solanum Quitoense* Lam.) fruit: Sensory (taste) versus biofunctional (ACE-inhibition) properties. Journal of Agricultural and Food Chemistry. 2016;**64**(26):5375-5383. DOI: 10.1021/acs.jafc.6b01631

[36] Hsu CL, Yen GC. Induction of cell apoptosis in 3T3-L1 pre-adipocytes by flavonoids is associated with their antioxidant activity. Molecular Nutrition & Food Research. 2006;**50**(11):1072-1079. DOI: 10.1002/mnfr.200600040

[37] González-Castejón M, Rodriguez-Casado A. Dietary phytochemicals and their potential effects on obesity: A review. Pharmacological Research. 2011;**64**(5):438-455. DOI: 10.1016/j.phrs.2011.07.004

[38] Yang JY, Della-Fera MA, Nelson-Dooley C, Baile CA. Molecular mechanisms of apoptosis induced by ajoene in 3T3-L1 adipocytes. Obesity. 2006;**14**(3):388-397. DOI: 10.1038/oby.2006.52

[39] Karim N, Jia Z, Zheng X, Cui S, Chen W. A recent review of citrus flavanone naringenin on metabolic diseases and its potential sources for high yield-production. Trends in Food Science and Technology. 2018;**79**:35-54. DOI: 10.1016/j.tifs.2018.06.012

[40] Yang M, Jiang ZH, Li CG, Zhu YJ, Li Z, Tang YZ, et al. Apigenin prevents metabolic syndrome in high-fructose diet-fed mice by Keap1-Nrf2 pathway. Biomedicine and Pharmacotherapy. 2018;**105**:1283-1290. DOI: 10.1016/j.biopha.2018.06.108

[41] Cho S, Choi Y, Park S, Park T. Carvacrol prevents diet-induced obesity by modulating gene expressions involved in adipogenesis

and inflammation in mice fed with high-fat diet. Journal of Nutritional Biochemistry. 2012;**23**(2):192-201. DOI: 10.1016/j.jnutbio.2010.11.016

[42] Son MJ, Rico CW, Nam SH, Kang MY. Effect of oryzanol and ferulic acid on the glucose metabolism of mice fed with a high-fat diet. Journal of Food Science. 2011;**76**(1):4-7. DOI: 10.1111/j.1750-3841.2010.01907.x

[43] Pongchaidecha A, Lailerd N, Boonprasert W, Chattipakorn N. Effects of curcuminoid supplement on cardiac autonomic status in high-fat-induced obese rats. Nutrition. 2009;**25**(7-8):870-878. DOI: 10.1016/j.nut.2009.02.001

[44] González-Ortiz M, Méndez-Del Villar M, Martínez-Abundis E, Ramírez-Rodríguez AM. Effect of resveratrol administration on metabolic syndrome, insulin sensitivity, and insulin secretion. Minerva Endocrinologica. 2018;**43**(3):229-235. DOI: 10.23736/S0391-1977.16.02550-5

[45] Rezazadeh K, Aliashrafi S, Asghari-Jafarabadi M, Ebrahimi-Mameghani M. Antioxidant response to artichoke leaf extract supplementation in metabolic syndrome: A double-blind placebo-controlled randomized clinical trial. Clinical Nutrition. 2018;**37**(3):790-796. DOI: 10.1016/j.clnu.2017.03.017

[46] Nishihira J, Sato-Ueshima M, Kitadate K, Wakame K, Fujii H. Amelioration of abdominal obesity by low-molecular-weight polyphenol (Oligonol) from lychee. Journal of Functional Foods. 2009;**1**(4):341-348. DOI: 10.1016/j.jff.2009.09.002

[47] González-Ortiz M, Martínez-Abundis E, Espinel-Bermúdez MC, Pérez-Rubio KG. Effect of pomegranate juice on insulin secretion and sensitivity in patients with obesity. Annals of Nutrition and Metabolism. 2011;**58**(3):220-223. DOI: 10.1159/000330116

[48] Ibero-Baraibar I, Azqueta A, De Cerain AL, Martinez JA, Zulet MA. Assessment of DNA damage using comet assay in middle-aged overweight/obese subjects after following a hypocaloric diet supplemented with cocoa extract. Mutagenesis. 2015;**30**(1):139-146. DOI: 10.1093/mutage/geu056

[49] Amin F, Islam N, Anila N, Gilani AH. Clinical efficacy of the co-administration of turmeric and black seeds (Kalongi) in metabolic syndrome-a double blind randomized controlled trial-TAK-MetS trial. Complementary Therapies in Medicine. 2015;**23**:165-174. DOI: 10.1016/j.ctim.2015.01.008

[50] FAO/WHO. Working Group Report on Drafting Guidelines for the Evaluation of Probiotics in Food. London, Ontario, Canada; 2002

[51] Lee E, Jung SR, Lee SY, Lee NK, Paik HD, Lim SI. *Lactobacillus plantarum* strain Ln4 attenuates diet-induced obesity, insulin resistance, and changes in hepatic mRNA levels associated with glucose and lipid metabolism. Nutrients. 2018;**10**(5):643. DOI: 10.3390/nu10050643

[52] Chen LH, Chen YH, Cheng KC, Chien TY, Chan CH, Tsao SP, et al. Antiobesity effect of *Lactobacillus reuteri* 263 associated with energy metabolism remodeling of white adipose tissue in high-energy-diet-fed rats. The Journal of Nutritional Biochemistry. 2018;**54**:87-94. DOI: 10.1016/j.jnutbio.2017.11.004

[53] Wang LC, Pan TM, Tsai TY. Lactic acid bacteria-fermented product of green tea and Houttuynia cordata leaves exerts anti-adipogenic and anti-obesity effects. Journal of Food and Drug Analysis. 2018;**26**(3):973-984. DOI: 10.1016/j.jfda.2017.11.009

[54] Balolong MP, Bautista RLS, Encarma NCA, Balolong EC Jr, Hallare AV, Elegado F. Evaluating the anti-obesity potential of *Lactobacillus fermentum* 4B1, a probiotic strain isolated from balao-balao, a traditional Philippine fermented food. International Food Research Journal. 2017;**24**(2):819-824

[55] Hsieh FC, Lan CCE, Huang TY, Chen KW, Chai CY, Chen WT, et al. Heat-killed and live *Lactobacillus reuteri* GMNL-263 exhibit similar effects on improving metabolic functions in high-fat diet-induced obese rats. Food & Function. 2016;**7**(5):2374-2388. DOI: 10.1039/c5fo01396h

[56] Plovier H, Everard A, Druart C, Depommier C, Van Hul M, Geurts L, et al. A purified membrane protein from *Akkermansia muciniphila* or the pasteurized bacterium improves metabolism in obese and diabetic mice. Nature Medicine. 2017;**23**(1):107-113. DOI: 10.1038/nm.4236

[57] An HM, Park SY, Lee DK, Kim JR, Cha MK, Lee SW, et al. Antiobesity and lipid-lowering effects of *Bifidobacterium* spp. in high fat diet-induced obese rats. Lipids in Health and Disease. 2011;**10**:116. DOI: 10.1186/1476-511X-10-116

[58] Fåk F, Bäckhed F. *Lactobacillus reuteri* prevents diet-induced obesity, but not atherosclerosis, in a strain dependent fashion in *Apoe−/−* mice. PLoS One. 2012;**7**(10):e46837. DOI: 10.1371/journal.pone.0046837

[59] Zhao X, Higashikawa F, Noda M, Kawamura Y, Matoba Y, Kumagai T, et al. The obesity and fatty liver are reduced by plant-derived *Pediococcus pentosaceus* LP28 in high fat diet-induced obese mice. PLoS One. 2012;**7**(2):e30696. DOI: 10.1371/journal.pone.0030696

[60] Lee SJ, Bose S, Seo JG, Chung WS, Lim CY, Kim H. The effects of co-administration of probiotics with herbal medicine on obesity, metabolic endotoxemia and dysbiosis: A randomized double-blind controlled clinical trial. Clinical Nutrition. 2014;**33**(6):973-981. DOI: 10.1016/J.CLNU.2013.12.006

[61] Sanchez M, Darimont C, Drapeau V, Emady-Azar S, Lepage M, Rezzonico E, et al. Effect of *Lactobacillus rhamnosus* CGMCC1.3724 supplementation on weight loss and maintenance in obese men and women. British Journal of Nutrition. 2014;**111**(8):1507-1519. DOI: 10.1017/S0007114513003875

[62] Sharafedtinov KK, Plotnikova OA, Alexeeva R, Sentsova TB, Songisepp E, Stsepetova J, et al. Hypocaloric diet supplemented with probiotic cheese improves body mass index and blood pressure indices of obese hypertensive patients—A randomized double-blind placebo-controlled pilot study. Nutrition Journal. 2013;**12**:138. DOI: 10.1186/1475-2891-12-138

[63] Kadooka Y, Sato M, Imaizumi K, Ogawa A, Ikuyama K, Akai Y, et al. Regulation of abdominal adiposity by probiotics (*Lactobacillus gasseri* SBT2055) in adults with obese tendencies in a randomized controlled trial. European Journal of Clinical Nutrition. 2010;**64**(6):636-643. DOI: 10.1038/ejcn.2010.19

[64] Omar JM, Chan YM, Jones ML, Prakash S, Jones PJH. *Lactobacillus fermentum* and *Lactobacillus amylovorus* as probiotics alter body adiposity and gut microflora in healthy persons. Journal of Functional Foods. 2013;**5**(1):116-123. DOI: 10.1016/j.jff.2012.09.001

[65] Kobyliak N, Conte C, Cammarota G, Haley AP, Styriak I, Gaspar G, et al. Probiotics in prevention and treatment of obesity: A critical view. Nutrition and Metabolism. 2016;**13**:14. DOI: 10.1186/s12986-016-0067-0

[66] Ghaisas S, Maher J, Kanthasamy A. Gut microbiome in health and disease: Linking the microbiome-gut-brain axis and environmental factors in the pathogenesis of systemic and neurodegenerative diseases. Pharmacology & Therapeutics. 2016;**158**:52-62. DOI: 10.1016/j.pharmthera.2015.11.012

[67] Le Chatelier E, Nielsen T, Qin J, et al. Richness of human gut microbiome correlates with metabolic markers. Nature. 2013;**500**(7464):541-546. DOI: 10.1038/nature12506

[68] Lehnen TE, Ramos da Silva M, Camacho A, Marcadenti A, Lehnen AM. A review on effects of conjugated linoleic fatty acid (CLA) upon body composition and energetic metabolism. Journal of the International Society of Sports Nutrition. 2015;**12**:36. DOI: 10.1186/s12970-015-0097-4

[69] Patterson E, Ryan PM, Cryan JF, Dinan TG, Ross P, Fitzgerald GF, et al. Gut microbiota, obesity and diabetes. Postgraduate Medical Journal. 2016;**92**(1087):286-300. DOI: 10.1136/postgradmedj-2015-133285

[70] Zarrati M, Salehi E, Nourijelyani K, Mofid V, Zadeh MJH, Najafi F, et al. Effects of probiotic yogurt on fat distribution and gene expression of proinflammatory factors in peripheral blood mononuclear cells in overweight and obese people with or without weight-loss diet. Journal of the American College of Nutrition. 2014;**33**(6):1-9. DOI: 10.1080/07315724.2013.874937

[71] Igual M, Ramires S, Mosquera LH, Martínez-Navarrete N. Optimization of spray drying conditions for lulo (*Solanum quitoense* L.) pulp. Powder Technology. 2014;**256**:233-238. DOI: 10.1016/j.powtec.2014.02.003

[72] Ministerio de Agricultura y Desarrollo Rural de Colombia. Producción Nacional por Producto. Available from: http://www.agronet.gov.co/Paginas/ProduccionNacionalProducto.aspx

[73] Contreras-Calderón J, Calderón-Jaimes L, Guerra-Hernández E, García-Villanova B. Antioxidant capacity, phenolic content and vitamin C in pulp, peel and seed from 24 exotic fruits from Colombia. Food Research International. 2011;**44**(7):2047-2053. DOI: 10.1016/j.foodres.2010.11.003

[74] Gancel AL, Alter P, Dhuique-Mayer C, Ruales J, Vaillant F. Identifying carotenoids and phenolic compounds in naranjilla (*Solanum quitoense* Lam. Var. puyo hybrid), an Andean fruit. Journal of Agricultural and Food Chemistry. 2008;**56**(24):11890-11899. DOI: 10.1021/jf801515p

[75] Acosta Ó, Pérez AM, Vaillant F. Chemical characterization, antioxidant properties, and volatile constituents of Naranjilla (*Solanum quitoense* Lam.) cultivated in Costa Rica. Archivos Latinoamericanos de Nutrición. 2009;**59**(1):88-94

[76] González-Loaiza DI, Ordóñez-Santos LE, Venegas-Mahecha P, Vásquez-Amariles HD. Cambios en las propiedades fisicoquímicas de frutos de lulo (*Solanum quitoense* Lam.) cosechados en tres grados de madurez. Acta Agronómica. 2014;**63**(1):11-17. DOI: 10.15446/acag.v63n1.31717

[77] Vasco C, Ruales J, Kamal-Eldin A. Total phenolic compounds and antioxidant capacities of major fruits from Ecuador. Food Chemistry. 2008;**111**(4):816-823. DOI: 10.1016/j.foodchem.2008.04.054

[78] Nutrition Facts of Lulo Fruit. Available from: https://www.traditionaloven.com/foods/details/fruits-juice/naranjilla-lulo-pulp-frozen-unsweeten.html

[79] Mertz C, Gancel AL, Gunata Z, Alter P, Dhuique-Mayer C, Vaillant F, et al. Phenolic compounds, carotenoids and antioxidant activity of three tropical fruits. Journal of Food Composition and Analysis. 2009;**22**(5):381-387. DOI: 10.1016/j.jfca.2008.06.008

[80] Van Berleere M, Dauchet L. Fruits, vegetables, and health: Evidence from meta-analyses of prospective epidemiological studies. In: Mariotti F, editor. Vegetarian and Plant-Based Diets in Health and Disease Prevention. London: Academic Press; 2017. pp. 215-248. DOI: 10.1016/B978-0-12-803968-7.00013-7

[81] Yahia EM, García-Solís P, Maldonado-Celis ME. Contribution of fruits and vegetables to human nutrition and health. In: Yahia EM, editor. Postharvest Physiology and Biochemistry of Fruits and Vegetables. Duxford: Woodhead Publishing; 2019. pp. 19-45. DOI: 10.1016/B978-0-12-813278-4.00002-6

[82] Fito P, Andrés A, Chiralt A, Pardo P. Coupling of hydrodynamic mechanism and deformation relaxation phenomena during vacuum treatments in solid porous food-liquid systems. Journal of Food Engineering. 1996;**27**(3):229-240. DOI: 10.1016/0260-8774(95)00005-4

[83] Fito P, Chiralt A, Barat JM, Andrés A, Martínez-Monzó J, Martínez-Navarrete N. Vacuum impregnation for development of new dehydrated products. Journal of Food Engineering. 2001;**49**(4):297-302. DOI: 10.1016/S0260-8774(00)00226-0

[84] Rivera-Espinoza Y, Gallardo-Navarro Y. Non-dairy probiotic products. Food Microbiology. 2010;**27**(1):1-11. DOI: 10.1016/j.fm.2008.06.008

[85] Vijaya-Kumar B, Vijayendra SVN, Reddy OVS. Trends in dairy and non-dairy probiotic products—A review. Journal of Food Science and Technology. 2015;**52**(10):6112-6124. DOI: 10.1007/s13197-015-1795-2

[86] Rad AH, Torab R, Mortazavian AM, Mehrabany EV, Mehrabany LV. Can probiotics prevent or improve common cold and influenza? Nutrition. 2013;**29**(5):805-806. DOI: 10.1016/j.nut.2012.10.009

[87] Panghal A, Janghu S, Virkar K, Gat Y, Kumar V, Chhikara N. Potential non-dairy probiotic products—A healthy approach. Food Bioscience. 2018;**21**:80-89. DOI: 10.1016/j.fbio.2017.12.003

[88] Kraus A. Factors influencing the decisions to buy and consume functional food. British Food Journal. 2015;**117**(6):1622-1636. DOI: 10.1108/BFJ-08-2014-0301

[89] Panghal A, Virkar K, Kumar V, Dhull SB, Gat Y, Chhikara N. Development of probiotic beetroot drink. Current Research in Nutrition and Food Science Journal. 2017;**5**(3):257-262. DOI: 10.12944/CRNFSJ.5.3.10

[90] Corbo MR, Bevilacvqua A, Petruzzi L, Casanova FP, Sinigaglia M. Functional beverages: The emerging side of functional foods commercial trends, research, and health implications. Comprehensive Reviews in Food Science and Food Safety. 2014;**13**:1192-1206. DOI: 10.1111/1541-4337.12109

[91] Grand View Research. Available from: https://www.grandviewresearch.com/

[92] Lumina intelligence.com. Available from: https://www.lumina-intelligence.com/

[93] Global Market Insights. Available from: https://www.globalinsights.com/

Chapter 5

The *Bifidobacterim bifidum* (BIB2) Probiotic Increased Immune System Factors in Men Sprint Athletes

Ali Hossein Khani, Seyed Milad Mousavi Jazayeri, Elahe Ebrahimi and Ayoub Farhadi

Abstract

Foods supplemented with probiotics enhance athletes' immune system functions, improve body health and consequently decreases athlete's health maintenance costs. Probiotics improve immune system function against pathogens via affecting on innate immune system, humeral immunity and cytokines. The effects of consumption of Iranian probiotic *Bifidobactrium bifidum* (BIB2) on athletes' immune system functions were evaluated. The results showed studied immune system factors were significantly different between test and control groups, so that IgA, IgM, lymphocyte and monocytes percentage and CD4 measurements of test group were higher than control. The *Bifidobacterim bifidum* (BIB2) probiotic consumption can affect some immune system factors; therefore its ability to improved general health should be studied more.

Keywords: probiotics, CD4, IgA, IgM, monocyte, lymphocyte, sprint athletes, *Bifidobactrium bifidum* (BIB2)

1. Introduction

There is a general belief among elite athletes and their coaches that overtraining causes resistance to infection. Epidemiological studies report that symptoms of respiratory tract infection increases in 1–2 weeks after strenuous endurance competitions. The highest percentages of patients were athletes who exceeded their training threshold level that is associated with the training load [1, 2]. The biological balance of body organs improves the health of the host, improving performance and increasing power of the immune system [3]. Probiotics are a group of living microorganisms that improve health by improving biological balance when added to foods or consumed as supplements. These organisms increase immune system function and enhance host defense against harmful microorganisms. The benefits of probiotics such as reducing toxins, increase immunity and resistance to infection, produce vitamins and nutrients, organic acids, reduce allergic reactions, respiratory infections, reduce the symptoms of irritable bowel syndrome, arthritis, rheumatoid and modulating immune responses have been shown in many studies [4, 5]. Overwhelming exercise undertaken by athletes or military personnel diminishes

IntechOpen

immune system function and increase gastrointestinal complaints, as well as increasing the risk of disease and infections [6–8]. In many studies, reduced immunity after chronic fatigue has been seen in over trained athletes [9–11]. Nieman et al. (2000) showed that regular and continuous exercise enhances the strength of athletes' immune system; while undertaking heavy and alternate physical activities had the reverse impact. Also, some research shows that excessive exercise also damages the immune system is impaired [12, 13].

The humoral immunity in athletes is often studied by mucosal immunoglobulin measurements, especially the changes in the secretion of IgA and IgM from tissue was reported in sporting activities [13]. Due to reduction of immunoglobulins in sport activities (and increased risk of infection in the upper respiratory tract) investigating the IgA and IgM is important [14]. Because the antibodies secretion, the lymphocytes have a very important role in the immune system [15]. Monocytes are the largest cells in the bloodstream and involved in phagocytosis in the early stages of the immune response. Also, monocytes produce the cytokines that activate lymphocytes and consequently stimulate inflammation [16]. Immune system malfunctions might be caused by stress, sleep disorder, exercise, and negative energy balance. Hard exercises raise neutrophil count; however, they decrease lymphocyte count, natural killer cell activity (NK cell) via disturbance with oxidative burst, neutrophil function, immunoglobulin's level and antimicrobial proteins level in saliva.

The effects of probiotics bacteria in improving immune function and preventing disease have been shown in numerous studies [6, 17–19]. The *Lactobacillus* and *Bifidobacterium* produce bactericidal and bacteriostatic agents such as lactic acid that can prevent pathogenic bacteria growth. Therefore probiotics reduce tissue inflammation directly or with antagonistic effects on pathogens [20, 21]. It has been shown that probiotics stimulate interferon-A secretion and this leads to increasing host phagocytic capacity [22, 23]. Kotani et al. reported that consuming *Lactobacillus* rised salivary IgA secretion [24]. Therefore, in this study the effects of juice supplemented by *Bifidobacterium bifidum* probiotic (2×10^9 cfu/ml) on immune factors including IgA, IgM, lymphocytes, monocytes and CD4 cells count were in the men's sprint athletes was investigated. The results could play an important role for athletes to overcome many diseases and infections, especially respiratory infections.

Effects of fruit juices supplemented by *Bifidobacterium bifidum* (BIB2) probiotic ($2*10^9$ cfu/ml) was assessed on immune system function of sprint athletes. Hence, 56 male athletes aged 21 ± 3 years and average weights of 78 ± 5 kg which divided into 2 groups: the first group received a glass (200 cc) juices containing probiotic daily for 12 weeks and control group received only simple juice (placebo) without probiotic. All volunteers had general health and did not smoke. Also they were asked not to use any probiotics products and antibiotics during the survey. The athletes had three times exercise per week. Before experiment, the probiotic characteristics and testing process were explained for all participants. All athletes voluntary participated in the experiment.

1.1 Blood sample collection

First, prior to the tests, the blood samples were collected from athletes after fasting, then once a month 24 h after 100 meters running, 8 ml of blood were collected from both studied groups in EDTA tubes. The bloods were centrifuged at 5000 rpm for 15 min and serums were collected. The blood cell counting was done using cell counter device (BC-2000 Mindray).

1.2 The IgA and IgM measurements

The IgA and IgM were measured using ELISA methodology using human IgA and IgM ELISA Kit (ab137980 and ab137982) according to protocol suggested by manufacture (abcam Inc., USA) using ELISA reader (Biotech microplate reader ELX800).

1.3 The CD4 cells count measurements

Flow cytometry is a technique for counting microscopic particles. The CD4 cells count was measured by flow cytometry technique using BD FACSCalibur system and anti-CD4 monoclonal antibody and results were reported based on the percent of peripheral blood mononuclear cell (PBMC) in total suspension.

1.4 Statistical analysis

The obtained data were analyzed using Graphpad Prism (version 6.01) software with 0.05 significant levels.

2. Results and discussion

The results of present study showed that consuming probiotic supplemented fruit juice increased serum IgA level and can be boosted immune system activities (**Table 1**). According to results shown in chart 1, CD4 cells count increased after consuming probiotics (**Figure 2**, **Table 1**) and it probably affects immune system ability to defense against antigens.

Probiotics or their products could have antimicrobial activities or they can prevent colonization of pathogens [4]. They probably have adjuvant effect so they may be stimulate phagocytosis done by leucocytes or may be increase IgA and defensins secretion. They could be attached to gut immune receptors, thus they inhibit competitively pathogenic virus's or bacteria's attachment. In addition, competition occurs for earning foods and nutrition so with suitable colonization of probiotics, we can conquer pathogenic bacteria.

Indigenous bacteria are accepted to add to the immunological protection of the host by making a hindrance against colonization by pathogenic microbes. This hindrance can be upset by sickness and by utilization of antibiotics, in this way, permitting less demanding access of the host gut by pathogens. It is presently trusted

Time	IgA (cell/μl)	
	Test group	Control group
	Means ± SD*	Means ± SD
Before test	364.954 ± 25.98	356.234 ± 42.36
First month	398.065 ± 40.24	350.568 ± 36.78
Second month	431.365 ± 37.96	358.653 ± 42.63
P-value		<0.05

**Standard division.*

Table 1.
The CD4 means ± standard deviation in test and control group.

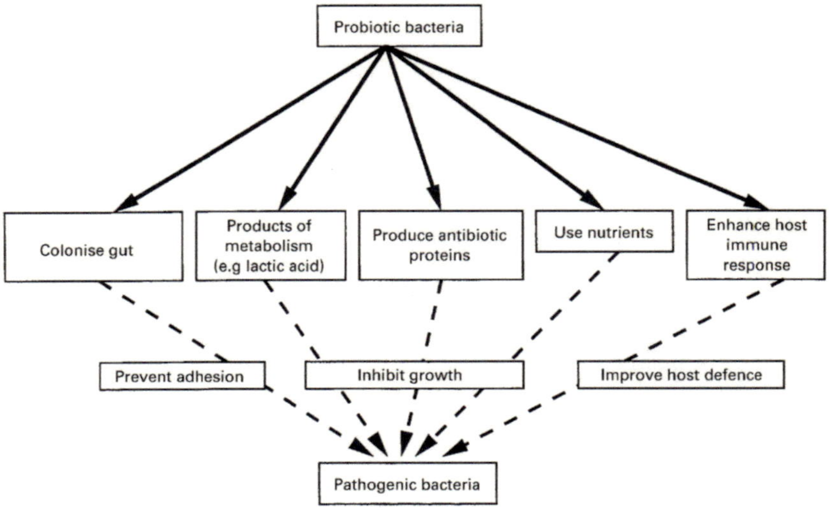

Figure 1.
Potential roles of probiotic bacteria in the human intestinal tract. Probiotic bacteria may act in a variety of ways to prevent the growth and colonization of pathogenic bacteria.

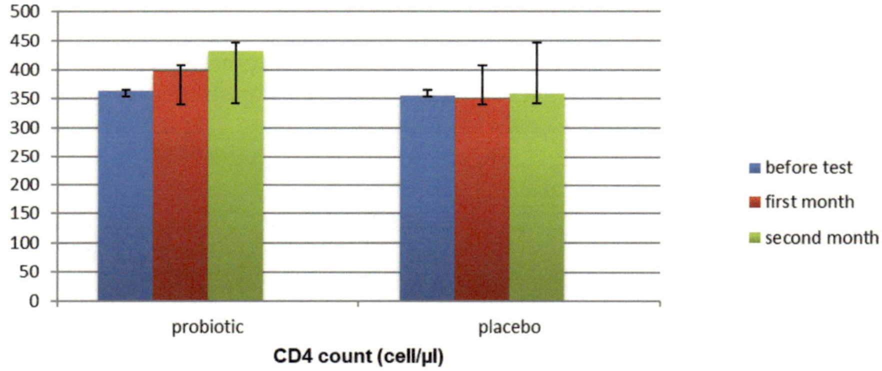

Figure 2.
The CD4 cells count (cell/µl) differences between test and control group in sprint athletes.

that this hindrance can be kept up by giving enhancements containing live 'alluring' microbes: such supplements are called probiotics [24].

n expansion to making a barrier impact, a few of the metabolic products of probiotic bacteria (e.g. lactic acid and a class of anti-microbial proteins named bacteriocins, created by a few bacteria) may hinder growth of pathogenic organisms. Moreover, the alluring bacteria may compete for nutrients with the pathogens. At last, there's a few prove that probiotic microbes may improve the intestine immune reaction against pathogenic microscopic organisms (**Figure 1**).

Studies in rats and mice uncover that lactic acid bacteria managed orally increment the numbers of T lymphocytes, CD4þ cells and antibody-secreting cells, counting those within the intestinal mucosa, and improve lymphocyte expansion, normal killer cell activity, IL-1, TNF and IFN-g generation, antibody production (counting secretory IgA), phagocytic activity and the respiratory burst of macrophages and the DTH reaction [21].

Hard and continues physical activities CAN decrease immune system function and it can lead to infectious diseases [25]. Probiotics have an important role to improve and boost an individual's health via pathogen growth prevention, amino acids and vitamins production, detoxification, cholesterol reduction and allergic reaction inhibition [26]. In the present study, the effects of *Bifidobactrium bifidum* (BIB2) probioticwe on immune system factors of IgA and CD4 were evaluated. Results revealed significant differences between the control and test group, so that swimmers received probiotic juice had higher IgA level and CD4 cells ($P < 0.05$) count than control group (**Tables 1** and **2**, **Figures 2** and **3**). These increment might accelerate microphage activities and phagocytosis, in turn, boost immune system against respiratory infections and diseases [27]. Lee et al. reported that intake of *Lactobacillus casei* and *Bifidobacterium* improved immune system functions and decreased respiratory infections among athletes [28].

West et al. reported that probiotic consumption lead to mucosal immune system improvement and also increased CD4 and dendritic cells [29]. In other research conducted by Ohashi et al. in the same field showed that increasing IgA level and cytokine secretion occurred during probiotic consumption [30]. Furthermore, it has been showed that probiotic consumption caused potent increasing in lymphocyte and NK cells count in peripheral blood samples and they improved immune system and general health while NK cells count reduction was determined during hard exercises [31, 32]. These results mentioned above were completely in accordance with our findings. Other studies have been done in the field of relationship between CD4 cells and probiotic consumption [33–35]. For example, Jensen et al. reported that consumption of *Bifidobactrium bifidum* probiotic increased CD4 cell, improved immune system functions and had anti-inflammatory effects [34]. Besides, findings obtained by Selbovitz et al. about the effects of probiotic on individuals suffered from acquired immune deficiency syndrome (AIDS) showed that probiotics could boost immune defense against viruses and could prevent virus transmission [35].

Lactobacillus and *Bifidobacterium* are normal inhabitants of the human adult gastrointestinal tract. Complex interactions occur between probiotic bacteria and the different constituents of the intestinal ecosystem (resident microflora and epithelial and immune cells). Mucosal epithelial surfaces, such as the gastrointestinal tract or respiratory tract that host a wide variety of different microorganisms from the external environment, are suitable sites for the onset of infection with pathogens. These levels are not protected. Different mechanisms of defense are involved in permanent and effective monitoring. Secretive secretion system plays an important role in this regard. IgA secretion (sIgA) is the dominant enzyme of the antibody in the secretion of the mammalian intestine. Most IgA is produced from suprapathic plasma cells that produce IgA polymer with the J chain (pIgA).

Time	IgA (μg/dl)		IgM (μg/dl)	
	Test group	Control group	Test group	Control group
	Means ± SD*	Means ± SD	Means ± SD	Means ± SD
Before test	230 ± 47	210 ± 56	143 ± 28	129 ± 73
First month	237 ± 48	185 ± 78	147 ± 75	132 ± 84
Second month	248 ± 75	193 ± 65	150 ± 62	135 ± 46
Third month	253 ± 83	194 ± 42	152 ± 43	133 ± 73
P-value				<0.05

**Standard division.*

Table 2.
The IgA and IgM means ± standard deviation in test and control group.

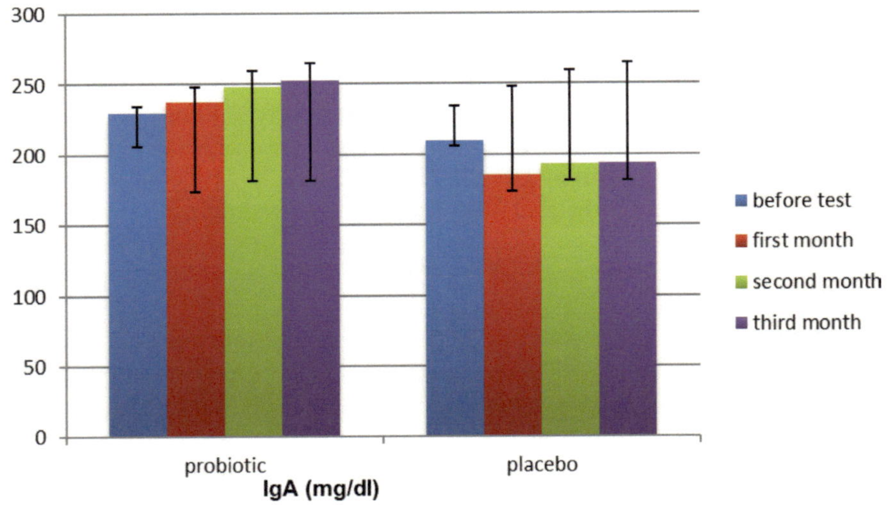

Figure 3.
The IgA concentration (mg/dl) differences between test and control group in sprint athletes.

Epithelial cells express the specific Ig receptor (pIgR). They are very important for the selective transfer of immunoglobulins to the lumen of the intestine. Immunoglobulin classes IgG and IgM are also present in the intestinal secretions, but vary in amounts and isotypes based on animal species. SIgA has many benefits. The dimer and tetrameric forms of IgA contain 4 to 8 antigen binding sites and have several "rewards" agents similar to IgM. IgA is more resistant to the activity of proteolytic enzymes that occur in gastrointestinal secretions [18]. IgA-antigen complexes do not activate the complement with inflammatory outcomes. Thanks to the content of mannan oligosaccharide side chains, sIgA could inhibit the adherence of bacteria with type I fimbriae to epithelial cells regardless of any specific antibody response. The typical response of the secretory immune system is the production of specific secretory IgA antibodies against luminal antigens to prevent other later responses on the epithelial surface. This process is called the immune exclusion and provides non-inflammatory protection in the mucous membrane. The integrity of the epithelial layer, the production of mucus, glycolipid, cytoprotein peptides and antibiotic-like agents are other host protection systems [29]. Protective micro-flora makes boundary impacts against risky pathogens and makes administrative specialists, such as brief and biofine fatty acids of bacteria. This impact moreover incorporates competition for receptors and metabolic foundations. It is more stamped but less caught on by the part of intestinal microflora within the modulation of homeostasis. The mucosal layer of the digestive tract is known as one of the biggest immune organs known to all sorts of immune cells [12]. It is conceivable to characterize anatomically the acceptance and effector parts of the immune reaction interior of the mucosal immune system. The most acceptance places are Peyer's patches restricted along the total little digestive tract. Lymphoid and accessory cells of Peyer's patches are secured by the follicular epithelium with M cells that serve as antigen preparing cells within the intestinal divider. Lymphocyte migration is important for the transport of immunological information between the different compartments of the intestinal immune system. The dendritic cells are the primary antigen particular cells depleting from the guts after mesenteric lymph node resection, afterward specially T-cells recycled through the intestine wall. After movement into the intestinal lamina propria, the lymphocytes may enter the space between the epithelial cells where they are present as intra-epithelial lymphocytes.

These intra-epithelial lymphocytes (IEL) may collectively constitute up to 27% of the epithelial cell populace and 40% of the peripheral T-cell populace. A tall extent of these cells is CD8+ (77% in pigs, 24% in sheep). They vary from blood T-cells. For case, they are CD90–, CD5– and carry an isoform of CD45 not found on peripheral blood T-cells. The division of the intestine wall from 5-day-old pigs come about in a 10-fold lower add up to lymphocyte surrender compared with grown-up pigs where 26.8×10^6 intra-epithelial lymphocytes and 35.2×10^6 add up to lymphocytes per g of tissue were gathered. Intestinal epithelial cells (IEC). They also contribute to the "education" of thymus independent subpopulations of intra-epithelial lymphocytes. Bacterial dependent activation of intestinal epithelial cells requires a direct contact with IEC and likely the interaction of surface molecules. For the start of the nearby resistant reaction and the actuation of particular T-cells, the entry of luminal antigens over the epithelial boundary is vital. Peyer's patches, or other lymphoid totals secured with a specialized epithelium layer with M cells are the most put for antigen section and T-cell actuation. The other sorts of immune cells such as macrophages, dendritic cells and enterocytes are moreover included within the handling of antigen at the mucosal level. T-lymphocytes from the intestinal lamina propria are ceaselessly beneath the antigen impact in vivo. They are actuated (IL-2 receptor expression, CD95) and viably respond to infection. Permanent antigen stimulation is responsible for the proliferation, maturation, and migration of T-cells to distant tissues where they act as effector cells in the immune response. T-cells produce lymphokines responsible for the aggregation of other types of immune cells (B cells, inflammatory cells) and for the modification of their microenvironment. One of the most important lymphokines is interferon γ (IFN-γ) produced by activated T-cells. It activates effector cells such as macrophages or epithelial cells (IEC) [29]. Murine IEC express MHC class II and ICAM-1 molecules and they display the antigen to T lymphocytes. This work is altered by the physiological or pathological status of the host. IEC are able to create in vitro a wide range of pro-inflammatory cytokines such as IL-8, MCP-1, TNF-α and GM-CSF in case they are affected by intrusive pathogenic bacteria. IL-8 and MCP are chemokines that pull in and actuate neutrophils and monocytes. TNF-α activate immune and inflammatory cells and GM-CSF incorporates a synergistic impact on cell actuation. The infiltration of the tissue with inflammatory effector cells to annihilate pathogens is continuously associated with a certain level of tissue harm. Subsequently the actuation is strictly controlled by external signals such as IFN-γ or TNF-α and by surface molecules like CD54 or CD95, which are able to enact or to discourage the actuation of IEC. There's a clear contrast between Gram-negative non-pathogenic microbes and lactic-acid bacteria (LAB) in their interaction with IEC. In coordinate interaction with IEC both types of bacteria actuate IFN-γ, but the stimulating impact of LAB is confined to the cellular surface molecule expression. The molecular mechanism responsible for these impacts of Gram-negative bacteria and LAB is not caught on. The capacity of Gram-negative bacteria to improve the expression of IFN-γ receptors on IEC increments the affectability of these cells to enactment with IFN-γ. On the other hand, LAB can stimulate IEC for successive actuation with this lymphokine, which is critical in local immune homeostasis. Permanent antigen stimulation of mucosal surfaces could create inflammatory lesions of the tissue. The homeostatic mechanisms should be active in the mucous layer to prohibit such undesirable effects. Apoptosis as programmed cell death shows one of the homeostatic mechanisms. Most of the T-lymphocytes in lamina propria are cells carrying a surface molecule – Fas that shifts apoptotic signals when it reacts with Fas ligand expressed on activated T-cells. Some T-cells also express Fas and Fas ligands that are potentially reactive with IEC or other T-cells [36].

Consumption of *Bifidobactrium bifidum* probiotic improves immune system defense ability against viruses such as influenza and enteroviruses [37, 38]. Also, these results were confirmed by Hu et al. [39]. Our findings indicated that consumption of *Bifidobactrium bifidum* (BIB2) probiotic increases CD4 cells count and they possibly improve the immune system response. According to results of present study and researches have been mentioned before, daily consumption of probiotics products could improve general health of individuals without any side effects. The results of this study show that probiotic juice containing *Bifidobactrium bifidum* probiotic supplementation affecting athletes sprint immune factors and showed significant differences with the control group. According to the survey was taken before the tests the participants feel a better perception of juice Probiotics than the Probiotics capsules.

In the present study athletes who consume probiotic showed the monocyte and lymphocyte cells significantly higher than control group (**Figures 4** and **5**, **Table 3**). Kekkonen et al. showed that in runners who take probiotics lower respiratory infections and gastrointestinal symptoms were reported, but the study did not find any effect of changes in monocyte cells that are inconsistent with our results [6].

West et al. reported the protective effect of probiotics on respiratory infections. In professional athletes who practice a lot in the long and intensive period, body temperature increased, the secretion of IgA confusion is created and therefore interferes in mucosal immunity [29]. Cox investigated the effect of probiotic *Lactobacillus fermentum* for 4 months on the Champ elite endurance. Unlike the results of this study did not observe any change in serum IgA, which can be because of the species and strains of bacteria [7]. Ashraf et al. showed that probiotics strengthen the immune system without causing an inflammatory response and immune modulators and directly affects the mucosal immune system. They suggested that probiotics as a natural and healthy food can increase stress resistance and immunity [40]. Probiotics can increase IgA antibody secretion, increase in phagocytic activity of macrophages and enhance the specific immune effects [41]. In study by Sakai et al. the effect of *Lactobacillus gasseri* on stimulating the production of IgA was shown [42].

Zhao and colleagues showed that *Lactobacillus plantarum* and *Bacillus subtilis* increase activity of lysozyme, superoxide (SOD) and the concentration of IgM

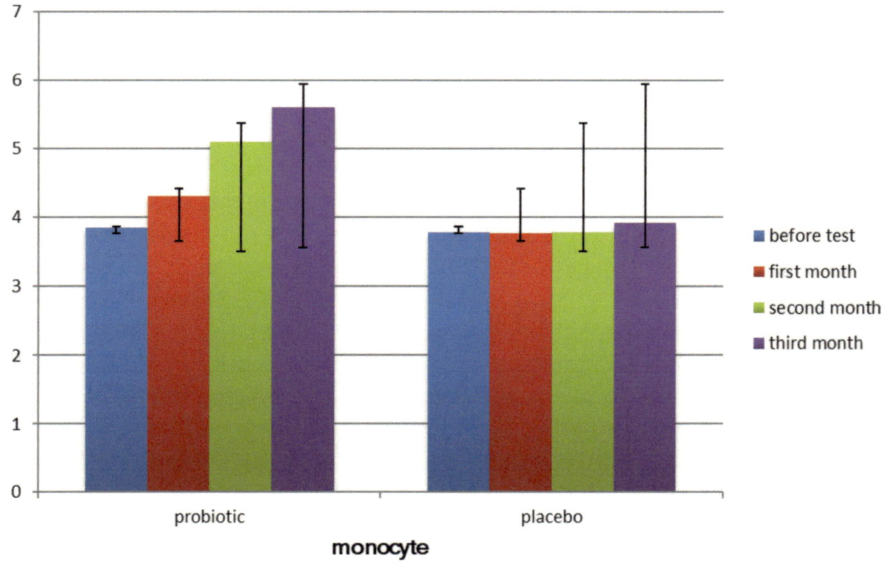

Figure 4.
The monocytes percentage differences between test and control group in sprint athletes.

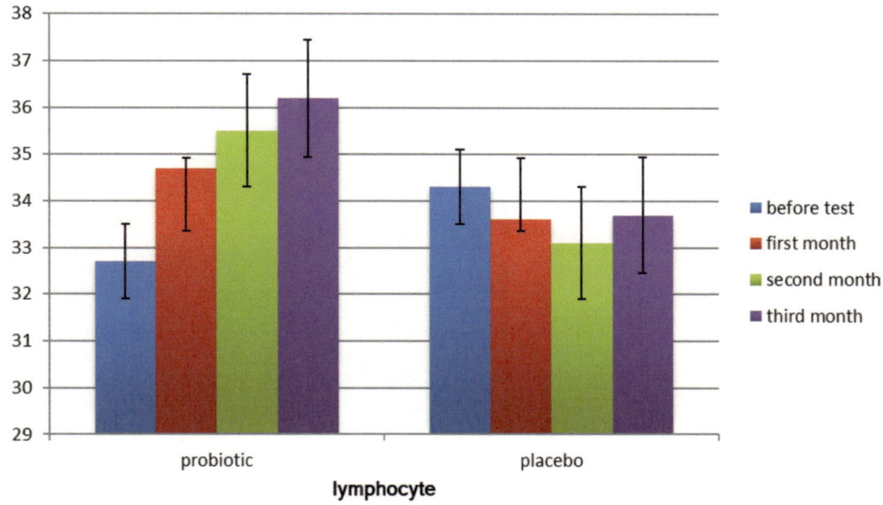

Figure 5.
The lymphocyte percentage differences between test and control group in sprint.

Time	Monocyte (%)		Lymphocyte (%)	
	Test group	Control group	Test group	Control group
	Means ± SD*	Means ± SD	Means ± SD	Means ± SD
Before test	3.85± 0.36	3.78 ± 0.42	32.7 ± 5.6	34.3 ± 4.8
First month	4.30± 0.65	3.77 ± 0.25	34.7 ± 6.4	33.6 ± 7.2
Second month	5.10± 0.42	3.78 ± 0.24	35.5 ± 5.7	33.1 ± 6.4
Third month	5.60± 1.2	3.91 ± 0.90	36.2 ± 7.3	33.7 ± 6.8
P-value				<0.05

**Standard division.*

Table 3.
The monocyte and lymphocytes means ± standard deviation in test and control group.

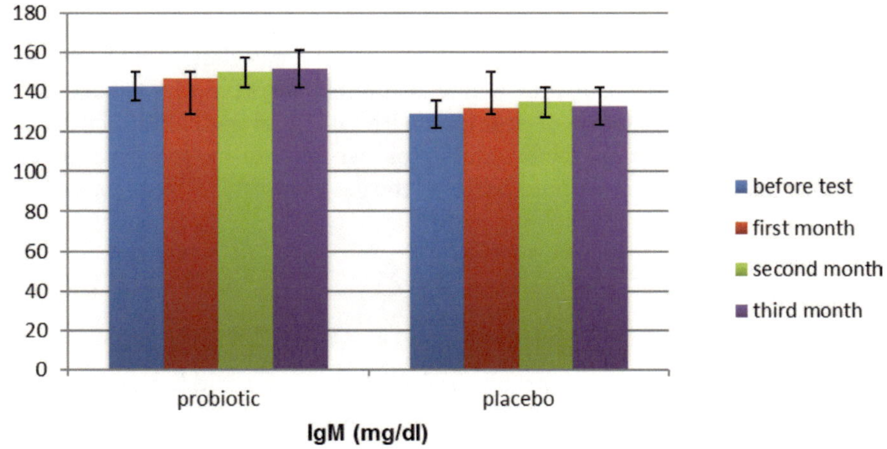

Figure 6.
The IgM concentration (mg/dl) count differences between test and control group in sprint athletes.

Time	Monocyte (%)	
	Test group	Control group
	Means ± SD*	Means ± SD
Before test	3.85 ± 0.36**	3.78 ± 0.42*
First month	4.30 ± 0.65*	3.77 ± .025*
Second month	5.01 ± 0.42*	3.78 ± 0.24*
Third month	5.60 ± 1.2	3.91 ± 0.90*
P-value		<0.05

**Standard division.*
***Means with same superscript letters in each row are significantly (P < 0.05) different.*

Table 4.
Comparsion the means of monocyte percentage between test and control group (n = 3) in sprint athletes.

Time	Lymphocyte (%)	
	Test group	Control group
	Means ± SD*	Means ± SD
Before test	32.7 ± 5.6a,**	34.3 ± 4.8a
First month	37.4 ± 6.4a	33.6 ± 7.2b
Second month	35.5 ± 5.7a	33.1 ± 6.4b
Third month	36.2 ± 7.3a	33.7 ± 6.8b
P-value		<0.05

**Standard division.*
***Means with same superscript letters in each row are significantly (P < 0.05) different.*

Table 5.
Comparsion the means of lymphocyte percentage between test and control group (n = 3) in sprint athletes.

that was consistent with our results [43]. In the present study in the athletes who drank juice containing the probiotic, the IgM was significantly higher than control group (**Figure 6** and **Table 2**). Probiotic bacteria in reaction with macrophage cells in tight junctions of epithelial cells, immune cells and dendritic cells led to the development of immune function, then macrophages and dendritic cells also generate immune responses through it. The dendritic cells spread their teeth between intestinal epithelial cells in the intestinal wall and use the probiotics to regulate immune function. The reaction of probiotic with intestinal epithelial cells induces the secretion of antimicrobial factors and cytokines and leading to activation of B and T lymphocytes in the lymphoid tissue of the gastrointestinal tract [6, 44, 32] (**Tables 4** and 5).

3. Conclusion

The advantageous impacts of probiotics have been illustrated in numerous diseases. One of the major mechanisms of probiotic activity is through the control of resistant reaction. A few of the prevalently utilized probiotic microorganisms are *Lactobacillus rhamnosus*, *Lactobacillus reuteri*, bifidobacteria and certain strains of *Lactobacillus casei, Lactobacillus acidophilus*-group, *Bacillus coagulans*, *Escherichia coli* strain Nissle 1917, certain enterococci, particularly *Enterococcus faeciumSF68*, and the yeast *Saccharomyces boulardii*. Bacterial spore formers, generally of the class Bacillus dominate the scene. These probiotics are included to foods, especially fermented drain items, either separately or in combinations.

As shown below a number of mechanisms are thought to be associated with probiotic beneficial effects:

1. Production of inhibitory substances such as H_2O_2, bacteriocins, organic acids, and so on.

2. Blocking of adhesion sites for pathogenic bacteria.

3. Compete with and inhibit growth of potential pathogens,

4. Degradation of toxins as well as the blocking of toxin receptors,

5. Modulate inflammatory immune responses [45].

The results of this study show that the probiotic *Bifidobactrium bifidum* BIB2) juice supplementation can significantly affect some immune system factors including IgA, IgM, lymphocytes, monocytes and CD4 cell count in sprint athletes compared to control group.

Author details

Ali Hossein Khani[1*], Seyed Milad Mousavi Jazayeri[2], Elahe Ebrahimi[3] and Ayoub Farhadi[4]

1 Department of Cellular and Molecular Biology, Young Researchers and Elite Club, Central Tehran Branch, Islamic Azad University, Tehran, Iran

2 Department of Microbiology, Science and Research Branch, Islamic Azad University, Guilan, Iran

3 Department of Microbiology, Faculty of Advanced Sciences and Technology, Tehran Medical Sciences, Islamic Azad University, Tehran, Iran

4 Laboratory for Molecular Cytogenetic, Faculty of Animal Sciences and Fisheries, Sari Agricultural Sciences and Natural Resources University, Sari, Iran

*Address all correspondence to: h.khani.ali@gmail.com

References

[1] Margonis K, Fatouros IG, Jamurtas AZ, Nikolaidis MG, Douroudos I, Chatzinikolaou A, et al. Oxidative stress biomarkers responses to physical overtraining: Implications for diagnosis. Free Radical Biology and Medicine. 2007;**43**(6):901-910

[2] Fry AC, Schilling BK, Weiss LW, Chiu LZ. β_2-Adrenergic receptor downregulation and performance decrements during high-intensity resistance exercise overtraining. Journal of Applied Physiology. 2006;**101**(6):1664-1672

[3] Prasanna P, Grandison AS, Charalampopoulos D. Bifidobacteria in milk products: An overview of physiological and biochemical properties, exopolysaccharide production, selection criteria of milk products and health benefits. Food Research International. 2014;**55**:247-262

[4] Gareau MG, Sherman PM, Walker WA. Probiotics and the gut microbiota in intestinal health and disease. Nature Reviews Gastroenterology and Hepatology. 2010;**7**(9):503-514

[5] Sanders ME, Guarner F, Guerrant R, Holt PR, Quigley EM, Sartor RB, et al. An update on the use and investigation of probiotics in health and disease. Gut. 2013;**62**(5):787-796

[6] Kekkonen R, Vasankari T, Vuorimaa T, Ekblom T, Hintsa A, Kautiainen H, et al. The effect of probiotics on respiratory infections and gastrointestinal symptom episodes during the marathon training period in marathon runners. International Journal of Sport Nutrition and Exercise Metabolism. 2007;**17**(4):352-363

[7] Cox AJ, Pyne DB, Saunders PU, Fricker PA. Oral administration of the probiotic Lactobacillus fermentum VRI-003 and mucosal immunity in endurance athletes. British Journal of Sports Medicine. 2010;**44**(4):222-226

[8] Bishop N, Gleeson M. Acute and chronic effects of exercise on markers of mucosal. Immunity. 2009;**1**(14):4444-4456

[9] Svendsen IS, Hem E, Gleeson M. Effect of acute exercise and hypoxia on markers of systemic and mucosal immunity. European Journal of Applied Physiology. 2016:1-11

[10] Reid V, Gleeson M, Williams N, Clancy R. Clinical investigation of athletes with persistent fatigue and/or recurrent infections. British Journal of Sports Medicine. 2004;**38**(1):42-45

[11] Cox AJ, Gleeson M, Pyne DB, Callister R, Hopkins WG, Fricker PA. Clinical and laboratory evaluation of upper respiratory symptoms in elite athletes. Clinical Journal of Sport Medicine. 2008;**18**(5):438-445

[12] Nieman DC. Is infection risk linked to exercise workload? Medicine and Science in Sports and Exercise. 2000;**32**(7 Suppl):S406-S411

[13] Barret S, Storey A, Harrison M. The effect of exercise mode on salivary IgA secretion in high level triathletes. Journal of Science and Cycling. 2014;**3**(2):3

[14] Moreira A, Mortatti AL, Arruda AF, Freitas CG, de de Arruda M, Aoki MS. Salivary IgA response and upper respiratory tract infection symptoms during a 21-week competitive season in young soccer players. The Journal of Strength & Conditioning Research. 2014;**28**(2):467-473

[15] Barbeau ML. Transactional Distances in an Online Histology

Laboratory Course. Ontario, Canada: The University of Western Ontario; 2012

[16] Abbas AK, Lichtman AH, Pillai S. Cellular and molecular immunology. Elsevier Health Sciences. 2014

[17] Spence L, Brown WJ, Pyne DB, Nissen MD, Sloots TP, McCormack JG, et al. Incidence, etiology, and symptomatology of upper respiratory illness in elite athletes. Medicine and Science in Sports and Exercise. 2007;**39**(4):577

[18] Lenoir-Wijnkoop I, Sanders ME, Cabana MD, Caglar E, Corthier G, Rayes N, et al. Probiotic and prebiotic influence beyond the intestinal tract. Nutrition Reviews. 2007;**65**(11):469-489

[19] Wanke M, Szajewska H. Lack of an effect of Lactobacillus reuteri DSM 17938 in preventing nosocomial diarrhea in children: A randomized, double-blind, placebo-controlled trial. The Journal of Pediatrics. 2012;**161**(1):40-43. e1

[20] Servin AL. Antagonistic activities of lactobacilli and bifidobacteria against microbial pathogens. FEMS Microbiology Reviews. 2004;**28**(4):405

[21] Gourbeyre P, Denery S, Bodinier M. Probiotics, prebiotics, and synbiotics: Impact on the gut immune system and allergic reactions. Journal of Leukocyte Biology. 2011;**89**(5):685

[22] Biswas D, Wideman NE, O'Bryan CA, Muthaiyan A, Lingbeck JM, Crandall PG, et al. Pasteurized blueberry (*Vaccinium corymbosum*) juice inhibits growth of bacterial pathogens in milk but allows survival of probiotic bacteria. Journal of Food Safety. 2013;**32**(2):204

[23] Arunachalam KD. Role of bifidobacteria in nutrition, medicine and technology. Nutrition Research. 1999;**19**(10):1559

[24] Kotani Y, Shinkai S, Okamatsu H, Toba M, Ogawa K, Yoshida H, et al. Oral intake of *Lactobacillus pentosus* strain b 240 accelerates salivary immunoglobulin A secretion in the elderly: A randomized, placebo-controlled, double-blind trial. Immunity and Ageing. 2010;**7**:11

[25] Glaser R, Kiecolt-Glaser JK. Handbook of Human Stress and Immunity. The University of Michigan: Academic Press; 1994

[26] Gareau MG, Sherman PM, Walker WA. Probiotics and the gut microbiota in intestinal health and disease. Nature Reviews Gastroenterology and Hepatology. 2010;**7**(9):503

[27] Sharma R, Kapila R, Kapila S. Probiotics as anti-immunosenescence agents. Food Reviews International. 2013;**29**(2):201

[28] Lee J-W, Shin J-G, Kim EH, Kang HE, Yim IB, Kim JY, et al. Immunomodulatory and antitumor effects in vivo by the cytoplasmic fraction of *Lactobacillus casei* and *Bifidobacterium longum*. Journal of Veterinary Science. 2004;**5**(1):41

[29] West N, Pyne D, Peake J, Cripps A. Probiotics, immunity and exercise: A review. Exercise Immunology Review. 2009;**15**(107):e26

[30] Ohashi Y, Ushida K. Health-beneficial effects of probiotics: Its mode of action. Animal Science Journal. 2009;**80**(4):361

[31] Yamada Y, Aoyama A, Tocco G, Boskovic S, Nadazdin O, Alessandrini A, et al. Differential effects of denileukin diftitox IL-2 immunotoxin on NK and regulatory T cells in nonhuman primates. The Journal of Immunology. 2012;**188**(12):6063

[32] Golden-Mason L, Cox AL, Randall JA, Cheng L, Rosen HR.

Increased natural killer cell cytotoxicity and NKp30 expression protects against hepatitis C virus infection in high-risk individuals and inhibits replication in vitro. Hepatology. 2010;**52**(5):1581

[33] Mengheri E. Health, probiotics, and inflammation. Journal of Clinical Gastroenterology. 2008;**42**:S177

[34] Schmidt EGW, Claesson MH, Jensen SS, Ravn P, Kristensen NN. Antigen-presenting cells exposed to *Lactobacillus acidophilus* NCFM, *Bifidobacterium bifidum* BI-98, and BI-504 reduce regulatory T cell activity. Inflammatory Bowel Diseases. 2010;**16**(3):390

[35] Selbovitz M. A review of probiotics studies in HIV research suggests improved immunological presentation and preservation of viral host restrictive factors of TH17 in HIV patients. Retrovirology. 2012;**9**(1):1

[36] Herich R, Levkut M. Lactic acid bacteria, probiotics and immune system. Veterinarni Medicina-PRAHA. 2002;**47**(6):169-180

[37] Leyer GJ, Li S, Mubasher ME, Reifer C, Ouwehand AC. Probiotic effects on cold and influenza-like symptom incidence and duration in children. Pediatrics. 2009;**124**(2):e172

[38] Colbère-Garapin F, Martin-Latil S, Blondel B, Mousson L, Pelletier I, Autret A, et al. Prevention and treatment of enteric viral infections: Possible benefits of probiotic bacteria. Microbes and Infection. 2007;**9**(14):1623

[39] Hu X, Zhang H, Lu H, Qian G, Lv L, Zhang C, et al. The effect of probiotic treatment on patients infected with the H7N9 influenza virus. PLoS One. 2016;**11**(3):e0151976

[40] Ashraf R, Shah NP. Immune system stimulation by probiotic microorganisms. Critical Reviews in Food Science and Nutrition. 2014;**54**(7):938-956

[41] Jespersen L, Tarnow I, Eskesen D, Morberg CM, Michelsen B, Bügel S, et al. Effect of *Lactobacillus paracasei* subsp. paracasei, *L. casei* 431 on immune response to influenza vaccination and upper respiratory tract infections in healthy adult volunteers: A randomized, double-blind, placebo-controlled, parallel-group study. The American Journal of Clinical Nutrition. 2015;**101**(6):1188-1196

[42] Sakai F, Hosoya T, Ono-Ohmachi A, Ukibe K, Ogawa A, Moriya T, et al. *Lactobacillus gasseri* SBT2055 induces TGF-β expression in dendritic cells and activates TLR2 signal to produce IgA in the small intestine. PLoS One. 2014;**9**(8):e105370

[43] Gao Q, Xiao C, Min M, Zhang C, Peng S, Shi Z. Effects of probiotics dietary supplementation on growth performance, innate immunity and digestive enzymes of silver pomfret, *Pampus argenteus*. Indian Journal of Animal Research. 2016;(OF)

[44] Gleeson M. Immune function in sport and exercise. Journal of Applied Physiology. 2007;**103**(2):693-699

[45] Narayan SS, Jalgaonkar S, Shahani S, Kulkarni VN. Probiotics: Current trends in the treatment of diarrhoea. Hong Kong Medical Journal. 2010;**16**(3):213-218

Chapter 6

Nutraceuticals from Microbes of Marine Sources

Charu Gupta and Dhan Prakash

Abstract

Therapeutic compounds can be derived from various natural sources like plants, animals, marine organisms, and microorganisms. Although the marine biota accounts for around 50% of the total world biodiversity, but their potential as a rich source of bioactive products and their applications in both pharmaceutical and nutraceutical industries have only recently been identified through several scientific studies. Marine biotechnology is an upcoming area that involves about the study of marine microorganisms and animals including algae, sponges, and coral as a novel source of bioactive substances that can be used in the treatment of various human diseases like cancer, anemia, diarrhea, obesity, diabetes, atopic dermatitis, Crohn's disease, etc. They are also potential sources of natural antioxidants, colors, immunosuppressants, enzyme inhibitors, hypocholesterolemic agents, vitamins, enzymes, and antibiotics. However, marine microorganisms have not yet been given the attention they deserve and a very limited scientific data is available on bioactive potential of marine microorganisms. There is still scope for a higher magnitude of research and investigation to explore the potential of both marine organisms and marine microorganisms as producers of novel drugs. This chapter deals with the exploitation of microbes from marine sources as potential sources for various nutraceuticals and their possibilities for applications in variety of diseases and as functional food supplement.

Keywords: microbes, nutraceuticals, marine organisms, functional food, bioactive compounds

1. Introduction

It is well-known that more than 70% of our planet's surface is covered by oceans. Experts estimate that the biological diversity in marine environment is higher than in tropical rain forests. Marine water contains enormous amounts of biodiversity which makes it as a source of huge amounts and wide varieties of novel bioactive compounds. The majority of the marine microbiota is soft bodied and follows a sedentary life style, thus requiring the other means of defense systems mainly by producing certain biochemical compounds that are generally toxic to the other animals. These toxic substances also aid them in detecting their harmful predators, and help them to protect themselves from their competitors or they can even paralyze their enemies. The biodiversity of marine microflora is overwhelming and there is an urgent need to explore and exploit their potential as biotherapeutic agents. These biotherapeutic compounds are usually synthesized as secondary metabolites by the marine microflora and fauna. The disadvantage is that since these natural products are synthesized and released extracellularly into the water so they are rapidly diluted and, therefore, their potency should be very high to show any effect. It is well-known that a large number of novel bioactive natural compounds are found in the oceans, and deep sea possessing various biological properties that can be exploited for discovering various drugs with improved efficacy and action in the treatment of various human diseases like cancer, anemia, diarrhea, obesity, diabetes, atopic dermatitis, Crohn's disease, etc. [1, 2].

The oceans are the source of a large group of structurally unique natural products that are mainly accumulated in invertebrates such as sponges, tunicates, bryozoans, and mollusks.

Macroalgae (or seaweed) are one of the most well-known types of algae used in the production of various nutraceuticals or dietary supplements. They are a rich source of valuable bioactive substances with both therapeutic and preventative effect. They have been used to develop a great variety of food and food ingredients, especially in Asian countries including Korea, Japan, and China. It is estimated that China and Indonesia are by far the largest seaweed producers with over 23 million tons of aggregated production in 2014 [3]. About 2400 natural products have been isolated from macroalgae belonging to the classes Rhodophyceae, Phaeophyceae, and Chlorophyceae [4]. Presently, seaweeds constitute commercially important marine renewable resources which are providing valuable ideas for the development of new drugs against diabetes, microbial infections, and inflammations [5]. Algal constituents include acids, alkaloids, amines, antibacterial, antifungal, antiviral substances, lipids, sterols, steroids, fatty acids, phenolic compounds, phytochromes, pigments, proteins, peptides, amino acids, sugar, alcohols, and vitamins. *Gracilaria opuntia* belongs to the family Rhodophyceae (Red algae) and possesses various biological activities.

The potential of marine microorganisms in producing various bioactive metabolites is due to their unique biochemical and physicochemical properties inherited by them in order to survive in extreme environmental conditions in the marine environment. These unique bioactive "bioceuticals" have tremendous potential for use as active pharmaceutical ingredient (API) and in food supplements to design various nutraceuticals [6].

Most of the bioactive compounds are having numerous biological activities which are found to act as nutraceuticals for humans and animals. Thus, the marine microflora contributes an important source of various bioactive constituents. Due to the great diversity of the marine flora, the chemistry of its associated bioactive compounds is also novel [7]. The marine flora includes a wide range of organisms from sponges, tunicates, bryozoans, mollusks to bacteria, microalgae, macroalgae, and cyanobacteria. The bioactive metabolites produced as a result of their metabolic activity is therefore effective in treatment of both infectious and non-infectious diseases.

Marine microflora including algae are widely used in the development of various nutraceuticals and are also used as food source or food ingredients [8]. Algae include both the micro- and macroforms and both the types are used as nutraceuticals. Microalgae are the most primitive and simply organized algae present in marine environment. They are the rich sources of various food nutrients and vitamins such as beta-carotene (vitamin A), vitamin C, E, H, B_1, B_2, B_6, and B_{12}, astaxanthin, polysaccharides, and polyunsaturated fatty acids [9]. These bioactive compounds are extracted from the microalgae and are used as food additives, fortifying infant milk, and other dietary supplements [10]. These compounds (especially the tunicate metabolite ET-743) have also shown good pharmacological properties and can be used to develop new drugs for cancer treatment. Some other compounds such as ziconotide obtained from the mollusk (*Conus magus*) are used as analgesics and anti-inflammatory. Natural products localized in symbiotic bacteria or cyanobacteria from marine invertebrates exhibits striking structural similarities with the known microbial metabolites; suggesting that bacteria and microalgae are involved in their biosynthesis and are the true sources of these metabolites. Nowadays, molecular techniques are used to study the microbial diversity in marine sponges and to study the involvement of bacteria in the biosynthesis of the bryostatins in the bryozoan *Bugula neritina* [2].

2. Marine microorganisms as source of nutraceuticals

A study reported that some unidentified prokaryotic communities, such as the JS1 and DSAG groups, occur widely in organic rich deep marine sediments associated with methane hydrates along the Pacific Ocean margin. The microflora is present in deep marine sediments and their community structure is affected by the surrounding geochemical and geological settings. Various studies have shown that many classes of microorganisms exist only in the sea [11]. Thus, limited scientific data is available on the growth media and culture techniques for culturing these marine microbes.

Many pharmaceutical industries have not been able to fully utilize this important resource. There is a general belief that marine microorganisms are difficult to culture; however, now there are a number of reports that showed that these marine microorganisms can be successfully cultured [12]. Thus, now many of the developed and underdeveloped countries have shifted their research focus on the marine habitat and new marine-oriented projects are emerging worldwide. Majority of microbes belonging to class bacteria and fungi are now the target of biomedical study. The coastal bacterial samples that grow under saline conditions are a source of novel antibiotics, antitumor, and anti-inflammatory compounds [13]. The symbiotic microbial consortia have also been proven to be a rich source of bioactive compounds with pharmaceutical potential. Many bacteria and fungi have been sampled from the surfaces of marine plants and the internal tissues of invertebrates, and they have been found to be of increasing interest [14].

2.1 Marine bacteria

Marine bacteria are prolific producers of valuable secondary metabolites as they thrive in harsh oceanic climates. The marine isolate *Pseudomonas*, Gram-negative, γ-proteo-bacteria is not well explored and only a limited number have been reported as producers of bioactive compounds. A bacterium strain KMM 3042 that is aerobic, non-pigmented, produces some bioactive substances such as pyrroles, pseudopeptide pyrrolidinedione, phloroglucinol, phenazine, benzaldehyde, quinoline, quinolone, phenanthren, phthalate, andrimid, moiramides, zafrin, and bushrin [15]. Some of these bioactive compounds are antimicrobial agents, and dibutyl phthalate and di-(2-ethylhexyl) phthalate have been reported to be cathepsin-B inhibitors [16].

Stenotrophomonas strains isolated from sponge, sea urchin, and ophiura specimens showed remarkable antimicrobial and antifungal inhibitory activity. However, they showed negligible activity against *Candida albicans*, but these strains could substantially inhibit Gram-positive microorganisms. Though *Stenotrophomonas maltophilia* is an opportunistic pathogen, it also possesses biocontrolling capabilities [17]. Low molecular weight antimicrobial metabolites have also been reported from marine ark shell *Anadara broughtoni* associated heterotrophic bacteria (butanol extracts) that exhibit strong antimicrobial, hemolytic, and surface activities [18]. Another recently discovered genus of bioactive substance producing marine bacteria is *Pseudoalteromonas*, the seawater species *P. phenolica* was reported to inhibit methicillin-resistant *Staphylococcus aureus* (MRSA) strains due to the presence of a brominated biphenyl compound, 3,3″,5,5″-tetrabromo-2,2″-diphenyldiol [19]. Some strains of *Pseudoalteromonas luteoviolacea* have also been shown to inhibit the growth of protists [20]. Other marine invertebrates have also been shown to be a source of novel bioactive compound which was later identified as a tambjamine (4-methoxypyrrole-containing bioactive compounds) like alkaloid and re-designated as YP1 [21]. *Pseudoalteromonas tunicate* is associated with them and produces a yellow pigment that possesses antifungal activity [20, 22]. These tambjamines are also known to possess antimicrobial, antitumorigenic, immunosuppressive, antiproliferative, and ichthyo-deterrent activities [20]. There are other evidences that points toward the colonizing bacteria present at the surface of higher organisms as the source of these compounds [23]. These studies have been further proven by Burke and colleagues by elucidation of YP1 biosynthetic pathway in *Pseudoalteromonas tunicata* [24].

2.2 Marine-derived fungi

Marine fungi have been known to produce a wide variety of bioactive metabolites that possess anticancer, antibacterial, antiplasmodial, anti-inflammatory, and antiviral activity [25, 26]. This is due to the presence of some unique and exceptional carbon frameworks in them. These novel compounds are used as new lead structures for medicine and for plant protection. The detailed procedure for their isolation and cultivation from various marine organisms (sponges, algae, and mangrove plants) has been given by Kjer et al. [27]. They have also elucidated the structure of secondary metabolites produced by these fungi. A novel anthraquinone derivative with naphtho [1,2,3-de]chromene-2,7-dione skeleton was isolated from a marine filamentous fungus, *Aspergillus glaucus* in the Fujian province of China and named aspergiolide A [28]. It was found to exhibit cytotoxicity against K562 and P388 cell lines. Similarly, *Penicillium* sp., isolated from deep ocean sediment was found to exhibit antitumor activities in their alkaloid-rich extracts due to the presence of meleagrin D and E and roquefortine H and I. However, they exhibited weak cytotoxicity in comparison to the previously reported meleagrin B and meleagrin that functions by inducing HL-60 cell apoptosis and can also arrest the cell cycle through G2/M phase, respectively. The mode of action was the distinct substitutions on the imidazole ring that have a significant influence on the cytotoxicity of these alkaloids [29]. Some other novel compounds and metabolites were also isolated and characterized from marine fungi *Ampelomyces* sp. that possesses potent antimicrobial and antifouling compounds. The antilarvicidal effect was due to the presence of compound 3-chloro-2,5-dihydroxybenzyl alcohol that effectively inhibited larval settlement of the tubeworm *Hydroides elegans* and of cyprids of the barnacle *Balanus amphitrite*. The compound is non-toxic and is also a potent antifoulant and/or antibiotic agent [30]. Another example of marine-derived fungus *Cladosporium* sp. also exhibited antibiotic and antifouling activity. It was later named as strain F14. The fungus produced the bioactive compounds in nutrient enriched cultivation media, in the presence of glucose or xylose [31]. Another study reported the marine-derived fungus *Fusarium* sp. (strain 05JANF165) to possess novel antimitotic and antifungal activity from its ethanol extracts. This compound was later identified and named as Fusarielin E [32].

Another marine-derived fungus of the genus *Pseudallescheria* was reported to produce a novel antibacterial dioxopiperazine, dehydroxybisdethiobis-methylthiogliotoxin from its broth. All three compounds exhibited potent antibacterial activity against the methicillin-resistant and multidrug-resistant *Staphylococcus aureus*, whereas Gliotoxin showed a significant radical scavenging activity against 1,1-diphenyl-2-picrylhydrazyl (DPPH) [33].

A marine isolate of the fungus *Exophiala* was also reported to exhibit a mild antibacterial activity against *Staphylococcus aureus* due to the presence of two novel antibacterial aspyrone derivatives, viz., Chlorohydroaspyrones A and B, and the previously described aspyrone, asperlactone, and penicillic acid from its broth [34].

Marine fungi also exhibited nematicidal effect along with antimicrobial activity. The same activity has also been reported from marine ascomycete *Lachnum papyraceum* (Karst.) [35].

Another marine-derived *Phoma herbarum* strain was reported with significant radical scavenging activity against DPPH due to the presence of halogenated benzoquinones (bromochlorogentisylquinones A and B) [36].

2.3 Marine-derived actinomycetes

Actinomycetes are well-known to be the producers of secondary metabolites. Many well-known antibiotics, such as streptomycin, erythromycin, and tetracycline, with potent biological activities are produced by them [37]. Many marine-derived actinomycetes were found to be the producers of novel antitumor [38], antimalarial [39], and antimicrobial compounds [40, 41]. Another marine-derived actinomycete namely *Nocardiopsis lucentensis* produced four novel 3-methyl-4-ethylideneproline-containing peptides called as Lucentamycins A-D from their fermentation broth [42]. Only compound Lucentamycins A and B exhibited strong *in vitro* cytotoxicity against HCT-116 human colon carcinoma [42]. In a similar study, marine-derived isolate of *Streptomyces* sp. were found to produce four new derivatives, Mansouramycin A-D, and the known 3-methyl-7-(methylamino)-5,8-isoquinolinedione from their ethyl acetate extract. These bioactive compounds, exhibited strong cytotoxicity with great degree of selectivity for non-small cell lung cancer, breast cancer, melanoma, and prostate cancer cells [43] suggesting their potential as anticancer drugs. Similarly, Perez and coworkers [44] isolated a macrodiolide Tartrolon D from the fermentation broths of *Streptomyces* sp. MDG-04-17-069. The isolated tartrolon was found to exhibit strong cytotoxic activity against three human tumor cell lines, viz., lung (A549), colon (HT29), and breast (MDA-MB-231) [44]. In yet another study, the secondary metabolites of a marine *Saccharomonospora* sp. yielded a novel alkaloid Lodopyridone, was found to be cytotoxic to HCT-116 human colon cancer cells [45].

Besides their antitumor and anticancerous potential, marine actinomycetes are also known for their antimicrobial activities also. A marine actinomycete, *Marinispora* was used to isolate a series of chlorinated bisindole pyrroles, Lynamicins A–E that exhibited a broad-spectrum antimicrobial activity against both the groups of Gram-positive and Gram-negative bacteria. These compounds were also effective against important drug-resistant pathogens, such as methicillin-resistant *Staphylococcus aureus* (MRSA) and vancomycin-resistant *Enterococcus faecium* [41]. In a similar study, Carlson et al. [46] isolated the two novel bioactive compounds namely dienoyl tetramic acids tirandamycin C and D from the marine environmental isolate *Streptomyces* sp. These compounds were effective against vancomycin-resistant *Enterococcus faecalis* and are structurally similar to the previously identified compounds Tirandamycins A and B with a slight variation in the pattern of pendant oxygenation on the bicyclic ketal system.

Similarly, ethyl acetate extract of *Streptomyces* sp. isolate B8652 was used to isolate Trioxacarcins A, B, and C along with three new derivatives designated as Trioxacarcins D, E, and F [47]. These types of trioxacarcins exhibited good antimicrobial, antitumor, and antimalarial activity. Similarly, the crude extract of a marine *Streptomyces* strain that was isolated from deep sea sediments, exhibited potent antifouling activity [48].

In yet another study, a marine-derived *Actinomyces* strain (NPS554) that was isolated from Japan yielded two trialkyl-substituted aromatic acids, Lorneic acid A and B. It was observed that Lorneic acid-A had significant inhibition activity against phosphodiesterase (PDE)5 [49]. PDE5 inhibitors are of great pharmacological importance as they are used in erectile dysfunctions and pulmonary hypertension.

2.4 Marine-derived microalgae

Cyanobacteria are a diverse group of Gram-Negative bacteria, also known as blue-green algae that produce an array of secondary metabolites with antifungal, antiviral, antibiotic, and other properties. They also exhibit selective bioactivity against vertebrates, invertebrates, plants, microalgae, fungi, bacteria, viruses, and cell lines [50]. Thus, they are of great pharmaceutical value. Besides, there are other anticancer compounds, which were initially thought to be obtained from marine sources, are now known to be produced by cyanobacteria [51]. Ulithiacyclamide and Patellamide A belong to Cyanobactins, produced by cyanobacteria; possess potent antimalarial, antitumor, and multidrug reversing activities [52].

Some other examples of marine cyanobacterial bioactive natural products are viridamides A and B. It was observed that Viridamide A produced by a blackish-green, mat-forming, filamentous cyanobacterium *Oscillatoria nigroviridis* showed antitrypanosomal, and antileishmanial activity [53]. In yet another study, the crude extracts of four green marine algae (*Cladophora rupestris*, *Codium fragile sp. tomentosoides*, *Ulva intestinalis*, and *Ulva lactuca*) were found to exhibit antiprotozoal activity [54]. All the algal extracts were active against *T. brucei rhodesiense*, and exhibited potent leishmanicidal activity [54]. This study was reportedly the first study to show antiprotozoal activity of British marine algae. Further, Desbois et al. [55] isolated an antibacterial polyunsaturated fatty acid, eicosapentaenoic acid (EPA) from the marine diatom, *Phaeodactylum tricornutum Bohlin*, which showed activity against a range of both Gram-positive and Gram-negative bacteria, including methicillin-resistant *Staphylococcus aureus* (MRSA) [55].

Both micro- and macroalgae are used as nutraceuticals. Studies have shown that microalgae are rich sources of all the vital nutrients such as beta-carotene, vitamins C, A, E, H, B_1, B_2, B_6, and B_{12}, astaxanthin, polysaccharides, and polyunsaturated fatty acids [9]. Thus, their bioactive molecules are produced commercially for use as food additives, infant milk formulations, and dietary supplements [10].

Macroalgae, are also commonly known as seaweeds. They are the most popular type of algae in the nutraceutical industry and are used in a great variety of food and food ingredients, especially in Asian countries like Korea, Japan, and China. Macroalgae are also well-known for the production of agarose. They are also an important source of many bioactive metabolites and natural products that possess many nutritional and therapeutic functions. Some of the examples of bioactive constituents are proteins, furanone, polyunsaturated fatty acids (PUFA), L-α kainic acid, phenotics, pigments, phlorotannins, phyco-colloids (carrageenan and agar), and minerals. Likewise, red and brown seaweeds are also good sources of many vitamins, minerals, proteins, and essential fatty acids [56, 57]. They are also used to prepare bioactive peptides and to improve protein digestibility. Further, antihypertensive bioactive peptides have also been isolated that can act as angiotensin-converting enzyme (ACE) inhibitors [58].

2.5 Symbiotic interaction between marine microbes

Symbiosis is the mutual association between any two organisms for their mutual benefit that can be in terms of their nutritional needs or protection from prey.

Research studies have shown that a variety of secondary metabolites are produced as a result of this association especially obtained from algae and invertebrates. Their associated microbes perform various biological activities [59]. Some of the various marine fungi isolated are *Haliclona simulans*, *Agaricomycotina*, *Mucoromycotina*, *Saccharomycotina*, and *Pezizomycotina* [60]. A variety of media were used for their isolation and identification and their antimicrobial activities were also determined. Some of these isolates exhibited antimicrobial activity against *Escherichia coli*, *Bacillus* sp., *Staphylococcus aureus*, and *Candida glabrata* [60]. It has been found that sponge-microbial association is a potential chemical and ecological phenomenon that can serve as a sustainable resource for generating novel pharmaceutical leads. Thus, sponge microsymbionts are an important focus nowadays [61, 62]. In yet another study, a marine-derived fungal strain named M-3 was isolated from marine red alga *Porphyra yezoensis.* It was assessed for its antifungal activity against *Pyricularia oryzae* [63]. As a result, a novel compound diketopiperazine was isolated from the culture extracts and its structure was also elucidated by spectroscopic methods.

In another study, the butanol extracts of algal associated species *Pseudoalteromonas issachenkonii* were reported to show hemolysis and inhibition of *Candida albicans*. Their ethyl acetate extracts were also subjected to spectroscopic studies and revealed the presence of indole-2,3-dione, a type of isatin that was responsible for its antifungal activity [63]. In another study, a red-brown hemolytic pigment was also discovered [64]. Coral reefs widely prevalent in oceans are also unexplored source of novel bioactive compounds [65].

In another study, a marine gorgonian associated bacterium *Bacillus amyloliquefaciens* was isolated from the South China Sea gorgonian, *Junceella juncea.* The studies showed their antibacterial against *Escherichia coli*, *Bacillus subtilis*, and *Staphyloccocus aureus* and antilarval properties against the larvae of bryozoan *Bugula neritina*. In yet another study, an antimicrobial activity of about 42 marine bacterial strains belonging to genera *Alteromonas*, *Pseudomonas*, *Bacillus*, and *Flavobacterium* was reported [66]. Besides this, certain mycoparasitic and fungicolous fungi were also shown to colonize other fungal physiological structures and were known to produce various bioactive agents [67]. In a study, five new natural products, Phomadecalins A, B, C, D, and Phomapentenone A, were reported from cultures of *Phoma* sp., a mitosporic fungal colonist isolated from the stromata of *Hypoxylon* sp. These bioactive compounds were found to be active against Gram-positive bacteria, *Bacillus subtilis* (ATCC 6051) and *Staphylococcus aureus* (ATCC 29213) [68].

3. Classification of marine nutraceuticals on the basis of chemical nature

Marine nutraceuticals can be broadly classified into Marine lipids (microalgal origin), polysaccharides derived from macro algae, marine probiotics, marine natural pigments, chitin and other related products, bioactive marine peptides/enzymes, and vitamins.

3.1 Lipids

Lipids derived from marine microalgae are used in larval nutrition of aquaculture, especially for enrichment of live feeds. Their other biological properties are anti-inflammatory, antiallergic, antiviral, and therapeutic. The wide spectrum of the properties is due to the presence of various components like polyunsaturated fatty acids (PUFA), highly unsaturated fatty acids (HUFA), and other substances. Various microalgal originated lipid/fatty acids and their activities [69] are given in **Table 1**.

Microalgal lipid/fatty acid	Biological action/function
Eicosapentaenoic acid (EPA)	Nutraceutical; antimicrobial and anti-inflammatory
Gamma-linolenic acid (GLA)	Integrity of tissue and delay of aging
Arachidonic acid (ARA)	Aggregative and vasoconstrictive of platelets
Docosahexaenoic acid (DHA)	Nutraceutical and brain development
Brassicasterol and stingmasterol	Hypercholesterolemic
Gamma-amino-butyric acid (GABA)	Neuro-transmitter, antioxidant and anti-inflammatory
Okadaic acid	Antifungal and promotion of the secretion of nerve growth factor (NGF)
Microcolin-A	Immunosuppressive

Table 1.
Microalgal lipid/fatty acids and their activity.

3.2 Polysaccharides

Generally, bacterial capsules contain polysaccharides. They form one of the important classes of secondary metabolites that are also important from pharmaceutical point of view. Research studies have shown that these exopolysaccharides (EPS) particularly from marine bacteria can be used in various pharmaceutical and food processing agents. They can also be used in industries as thickeners, coagulating agents, adhesive agents, stabilizers, and as gelling agents. The exopolysaccharides possess good viscosity and pseudo-plastic properties that impart them ability to resist extremes of temperature, pH, and salinity. This increases their potential to be used as an industry friendly resource [70]. In a study, exopolysaccharides have been shown to possess immunomodulatory and antiviral properties on immunocompetent cells in a marine bacteria *Geobacillus thermodenitrificans* that was isolated from a shallow marine vent of Volcano Island (Italy). This bacterium not only produced secondary metabolites against other organisms, but also produced certain compounds which help in bioremediation [71]. Certain other marine bacterial species are known as prolific producers of biosurfactants, bioemulsifiers, and exopolysaccharides.

Some polysaccharides are also derived from macroalgae. Seaweeds contain higher amounts of the polysaccharides like agar, alginates, and carrageenan. These act as food fiber and are collectively called phyco-colloids or hydrocolloids. Being rich in fiber, seaweeds exhibit health benefits like reducing the absorption of toxins, anticarcinogenic, and antioxidant properties. Example of the few important bioactive polysaccharides isolated from macroalgae (seaweeds) that possess the potential to be used as nutraceuticals are Fucoidan, Sphinganine amide, and caulerpicin (green algae), Carrageenan, Alginic acid and xylofucans, Hyperoxaluria, Sulfated polysaccharides, and Alginates. All these compounds possess wide range of biological properties such as antioxidant, antiangiogenic, antibacterial, antiviral and antitumor activities anticoagulant, immuno-modulating, hypolipidemic, and anti-inflammatory [72]. In addition to the phyco-colloids, seaweeds are sources of biologically active phytochemicals like carotenoids, phycobilins, fatty acids, vitamins, sterols, tocopherol, phycocyanins, and others.

Macroalgae are also rich sources of insoluble and soluble dietary fiber. They contain chiefly indigestible sulfated polysaccharides. Some of the notable examples of structural and storage polysaccharides are fucan, agar, laminaran, carrageenan, and alginates that are found both in red and brown seaweeds. The alginates obtained from brown seaweeds are used as hydrocolloids and fucans from brown seaweeds are used in both food and cosmetics industries [73].

A research highlighted the role of marine bacteria *Bacillus circulans* in the biodegradation of anthracene (a polyaromatic hydrocarbon) [74]. The said bacteria also produced a novel type of biosurfactant that exhibited excellent emulsification properties. It was shown that *Bacillus circulans* utilized anthracene as a sole carbon source for the production of biosurfactant. The researchers also reported the production of another biosurfactant by an unnamed marine bacteria that has the ability to remove metal from solutions [75].

3.3 Bryostatins: bryozoan origin

The marine bryozoan, *Bugula neritina*, is the sole source of the bryostatins, a family of macrocyclic lactones with anticancer activity. Bryostatins are actually the bacterial products as *B. neritina* harbors the uncultivated gamma proteobacterial symbiont *Candidatus Endobugula sertula*. The clinical studies of bryostatins are also under going to study their potential for the treatment of leukemias, lymphomas, melanomas, and solid tumors [76]. Their mode of action is that they act through protein kinase C signal transduction to alter cellular activity.

3.4 Probiotics: marine lactic acid bacteria (LAB) origin

Microbial diversity of marine environments is very rich and can be helpful to develop safe and effective probiotics. Novel marine probiotics can be an effective alternative for fighting the antibiotic resistance. *Lactobacillus* and *Bifidobacterium* are found to possess antimutagenic [77] and immunomodulatory [78] activity in host animal. Different strains of marine probiotic bacteria are *Lactobacillus* (*L. casei*, *L. acidophilus*, *L. rhamnosus* GG (ATCC 53013), *L. johnsonii* La-1), *Bifidobacterium* (*B. bifidum*, *B. longum*, *B. infantis*, *B. breve*, *B. adolescentis*), *Leuconostoc* spp. (*Ln. lactis*, *Ln. mesenteroides subsp. cremoris*, *Ln. mesenteroides subsp. dextranicum*), and *Streptococcus* spp. (*S. salivarius subsp. thermophiles*).

The problem posed during the development of new marine probiotics is the isolation and identification of potential strain. Application of biotechnological and molecular biological tactics is necessary for the development of marine probiotic strains for use of aquatic industry [72].

3.5 Pigments: marine algae

Besides polysaccharides and lipids, marine macro- and microalgae also provide various types of the bioactive natural pigments. The natural pigments of the marine algae provide food by photosynthesis and also provide the pigmentation. In addition to these, the natural pigments are also found to exhibit health benefits which make them one of the important marine nutraceuticals. Chlorophyll-a, Lutein, zeaxanthin, and canthaxanthin possess antimutagenic properties; pheophytin-a exhibits neuroprotective, and anti-inflammatory action; Chlorophyll-a, Pheophorbide-a, Pyropheophytin-a, Phycoerythrobilin, Lutein, Fucoxanthin, Phycocyanin, Astaxanthin, Zeaxanthin, and Canthaxanthin are all good antioxidants. Alpha-Carotene is also used as a food additive [72].

3.6 Chitosan: chitin

Chitosan is a natural polymer derived from chitin and it is the second most abundant polysaccharide after cellulose. Chitin is recovered from processing discards of shrimp, crab, lobster, and crayfish following de-proteinization and demineralization. The chitin so obtained may then be deacetylated to afford chitosan.

Fungal cell walls are also rich in chitin. Chitin and chitosan are used as biomaterials in a variety of application in edible film industry, additives, for improving nutritional quality, recovery of solid materials from food processing waste, in purification of water, etc. The other biological properties of chitosan are antioxidant, hypocholesterolemic, antimicrobial, and anti-inflammatory activities [79].

Chitosan disrupts the barrier properties of the outer membrane of Gram-negative bacteria due to ionic interaction between the cationic groups of the chitosan molecules and the anionic groups of the microbial cell membrane, which can rupture the cell membrane. Chitosan can also function as an antifungal agent by forming gas-permeable coats, interference with fungal growth and stimulation of various defense processes like, build-up of chitinases, production of proteinase inhibitors, and stimulators of callous synthesis.

The antioxidant property could be attributed to the ability of chitosan to chelate metals and combine with lipids. Derivatives of chitosan, namely, N,O-carboxymethyl chitosan, N,O-carboxymethyl chitosan lactate, N,O-carboxymethyl chitosan acetate, and N,O-carboxymethyl chitosan pyrrolidine carboxylate had also exhibited the antioxidant activity. Chitosan possesses special properties for use in pharmaceutical, biomedical, food industry, health, and agriculture due to its biocompatibility, biodegradability, and non-toxic nature. Through encapsulation, it is being used as a vehicle for nutraceutical compounds and pharmacological compounds. Chitosan derivatives may also be produced in order to obtain more effective products for certain applications [72].

3.7 Bioactive peptides/enzymes: marine origin

Peptides refer to the specific protein fragments that exhibit a specific biological activity. Some of the peptides may exhibit multifunctional properties like opioid, immunomodulatory, antibacterial, antithrombotic, and antihypertensive activity. Biofunctional peptides have a size range of 2–20 amino acid residues and are encrypted within the sequence of the parent protein and are released during processing. They can be formed either by acid or alkaline hydrolysis. The major bioactivities of peptides are antihypertensive (ACE inhibitory), antioxidant, antimicrobial, antihypoallergenic activity, and cell immunity [80].

Proteins isolated from bacteria such as *Dunaliella*, *Phaeodactylum tricornutum*, and *Arthrospira platensis* possess potent antioxidant and anti-inflammatory activity which can be effectively used in aquaculture practices. Similarly, enzymes such as superoxide dismutase and carbonic anhydrase derived from *Porphyridium*, *Anabaena*, *Isochrysis galbana*, and *Amphidinium carterae* play an important role in regulating the metabolite waste (CO_2).

3.8 Vitamins: marine microalgal origin

Marine microalgae are also known to have good amount of alpha-carotene. Microalgae like, *Arthrospira*, *Isochrysis galbana*, *Porphyridium cruentum*, and *Tetraselmis* are rich in vitamin C, K, A, E, and alpha-carotene which possess strong antioxidant activity. Vitamins A specially provitamin-A or alpha-carotene and vitamin E or alpha-tocopherol function as source of strong antioxidant compounds and protect the cells from free radical damage by quenching these free radicals. Vitamin E, together with vitamin C and alpha-carotene, helps in improving antioxidant defenses in the body. Fat soluble vitamin K isolated from Pavlova helps in blood clotting or coagulation. The role of antioxidant vitamins in health and disease control has been well documented [73]. Some of the important marine microorganisms, their bioactive metabolites and biological activity are highlighted in **Table 2** [20].

Microorganism	Marine microbial metabolites	Biological activity
Cyanobacteria	Dolastatin-10	Antimicrotubule; and the synthetic analog, TZT-1027, as antitumor
	Dolastatin-15	Antimicrotubule; and the synthetic analog, ILX-651, as antitumor
	Curacin A	Antimicrotubule
	Toyocamycin	Antifungal
Actinomycetes	Resistoflavine	Anticancerous and antibacterial
	Marinomycin A	Antitumor and antibiotic
	Daryamide C	Antitumor
	Violacein	Antiprotozoal
Bacteria	Macrolactin S	Antibacterial
	Pyrone I and II	Antibacterial
	MC21-B	Antibacterial
Fungi	Meleagrin	Antitumor
	Oxaline	Antitumor
	Alternaramide	Antibacterial
Algae	Norharman	Enzyme inhibitor
	Calothrixin-A	Antimalarial and anticancerous
	Eicosapentanoic acid (EPA)	Treats heart disease, anti-inflammatory agent (rheumatoid arthritis and immunodeficiency diseases)
Symbiotic microbes	Macrolactin V	Antibacterial and antilarval
	DAPG	Antibacterial (anti-MRSA, anti-VRSA and anti-VRE)
	BE-43472B	Antibacterial (anti-MRSA and anti-VRE)

Table 2.
Name and biological activity of some of the marine microbial metabolites.

4. Classical examples of some marine microflora and their nutraceutical potential

The oceans harbor one of the most diverse flora and fauna on our earth surface. Its biodiversity serves as an inexhaustible source of variety of biologically active compounds such as antibiotic, antimicrobial, anti-inflammatory, anticancer, antioxidant, antimicrotubule, cytotoxic, photo-protective, and antifouling properties. A variety of novel range of microorganisms, such as bacteria (both free living and symbiotic), fungi, actinomycetes, microalgae-cyanobacteria, and diatoms, are found in the marine environment that are potent producers of important therapeutic compounds. They have also been found effective against many deadly infectious diseases such as AIDS, drug-resistant bacteria, including conditions of multiple bacterial infections. Only little research has been done on the biophysical and biochemical properties, their chemical structures and biotechnological applications of these marine bioactive substances, and their potential utilization in both cosmeceuticals and nutraceuticals. Some of the research studies on bioactive molecules from marine sources are discussed below.

Seaweeds provide a rich source of bioactive molecules. In this study, the antimicrobial potential of Red sea weed, *Gracilaria opuntia* was investigated against clinically important microorganism such as *Escherichia coli*, *Staphylococcus aureus*, *Klebsiella pneumonia*, and *Pseudomonas putida* that cause diseases in diabetic patients. Crude extracts prepared from aqueous, ethanol, and methanol extraction procedures revealed that aqueous extraction procedure have a wide range of antimicrobial activity against all the test pathogens. The overall antibacterial activity assessed from the above results indicates the presence of active constituents in the extractions of seaweeds, which can be explored for the production of significant molecules that could be used in pharmaceutical industry [81].

In a similar another study on *Gracilaria edulis*, their phytochemical, antibacterial, and antifungal activities of crude extracts were investigated. The methanol and aqueous extracts of *Gracilaria edulis* showed the presence of a number of metabolites such as alkaloids, saponin, phenols, terpenoids, proteins, flavonoids, glycosides, coumarins, and tannins. The red algae showed the significant antibacterial activity against the clinical pathogens of *Pseudomonas aeroginosa*, *Staphylococcus aureus*, *Bacillus cereus*, *Bacillus subtilis*, *Streptococcus pyogenes*, *Salmonella typhi* as well as the fungus *Aspergillus niger*, *Aspergillus flavus*, *Rhizopus indicus*, and *Candida albicans*. The methanol extract showed the broader spectrum of antibacterial and antifungal activity when compared with aqueous extract [82].

In yet another study, the phytochemical and biological evaluation of some *Sargassum* species from Persian Gulf was also studied. These plants contain important phytochemical constituents and have various potential biological activities. The study investigated the presence of phytochemical constituents and total phenolic quantity of the seaweeds *Sargassum angustifolium*, *Sargassum oligocystum*, and *Sargassum boveanum*. Cytotoxicity and antioxidant potential of these three *Sargassum* species was also analyzed. *Sargassum angustifolium* had the highest content of total phenolics and showed the highest antioxidant activity. Cytotoxic results showed that all species could inhibit cell growth effectively, for *S. oligocystum*, *S. angustifolium*, and *S. boveanum*, respectively. Thus considerable phytochemicals and moderate cytotoxic activity of *S. angustifolium*, *S. oligocystum*, and *S. boveanum* make them appropriate candidate for further studies and identification of their bioactive principles [83].

Further study on screening the algae for their phytochemicals from their extracts and testing on their antibacterial, antifungal and antioxidant potential was carried out. The algal strains were *Tetraselmis* sp., *Dunaliella* sp., *Chlorella* sp., *Synechocystis* sp., and *Oscillatoria* sp. Their extracts were prepared in an organic solvent and were used to screen for the presence of any phytochemicals. Later, their antimicrobial activity was also assessed against some selected bacterial and fungal species. Their antioxidant activity was also determined by DPPH scavenging assay and confirmed the presence of flavonoids in majority of the solvent extracts. In addition, the microalgal strains exhibited better antifungal activity as compared to bacterial. The solvent acetone from *Dunaliella* sp. showed highest antioxidant activity. The results showed the presence of Octadecanoic acid-4-hydroxy-methyl ester, benzoic acid, hexadecanoic acid, and Tetradecanoic acid, confirming the bioactive compounds in the algal extracts [84].

5. Other marine sources of nutraceuticals

Besides the microbes (which contributes around 19%), there are other marine flora and fauna, such as sponges (38%), coelenterates (23%), algae (10%), echinoderms (7%), tunicates (7%), mollusks (3%), and bryozoans (2%), have also shown to exhibit their potential to produce various therapeutic compounds including some novel anticancer substances. These compounds can also work against infectious diseases and inflammation [85].

6. Current market of nutraceuticals

Nutraceuticals, including functional foods and dietary supplements, have tremendous market potential. It has been estimated that the consumer demand of these health foods were around $250 billion only in 2014 alone and this figure is continue to rise and is expected to reach around $385 billion by 2020 [86].

Nutraceutical products are in demand throughout the world especially in developed countries, including United States of America (USA), Europe, Japan, Asia Pacific, Middle East, and Latin America. In particular, the global market is dominated by the United States of America, Europe, and Japan, which account for more than 85% of the market [87]. As per the Mintel survey carried out in the United Kingdom on vitamins and minerals supplements, it was observed that 25% of all adult populations were satisfied with the results of the nutraceutical products. The percentage of the usage varied according to the age of the consumers like the consumption of nutraceutical products was more common in elderly population as compared with the younger generation. There is a strong belief that in the coming years, nutraceutical industry would remain at the forefront market including the Asian countries such as India and China. This is because of greater consumer awareness toward their health, increasing income levels and greater confidence in traditional and complimentary medicines [73].

The consumption of nutraceutical products in the USA is comparatively higher and accounts for approx. 40% of adults; in Spain, the nutraceuticals are consumed by the population aged between 35 and 80 years, and only around 9% consume dietary supplements as the source of vitamins and minerals. Besides, about 72% educated women that are of age between 35 and 49 years are more likely to choose nutraceuticals and dietary supplements [88].

In addition, the geriatric populations are also the most common consumers of these health supplements as they are more prone to micronutrient deficiencies. Due to the steady rise of geriatric population in developed countries, there is an urgent need to encourage and maintain a healthy lifespan and prevent chronic illnesses associated with aging [89].

Thus individuals with healthy lifestyles are more commonly the users of nutraceutical products.

7. Conclusion

From the above discussion, it is quite pertinent to conclude that the marine environment harbors variety of microbial flora that have capability to produce a wide array of bioactive metabolites that can be used both in nutrition and pharmaceuticals to formulate new drugs that are effective against various drug-resistant pathogens. Though the saga of marine microbial bioactive metabolites is continuing with new compounds being added day by day, our knowledge is still a miniscule of what exists deep in the oceans. Interdisciplinary research and collaborative endeavors are required amongst scientists, medical practitioners, marine microbiologists, and biotechnologists, to provide innovative approaches to marine based biomedical research. Thus, it is imperative to utilize our marine biodiversity and their bioactive metabolites for discovering new therapeutic compounds of nutraceutical importance. Recent biomedical tools, such as metabolomics and genetic engineering, can also be applied to increase their yield.

Acknowledgements

Authors greatly acknowledge Dept. of Scientific & Industrial Research (DSIR), New Delhi for their financial support. The authors are also grateful to Dr. Ashok K. Chauhan, Founder President and Mr. Atul Chauhan, Chancellor, Amity University-UP, Noida, India for the motivation and research facilities.

Conflict of interest

The author(s) confirm that this article content has no conflict of interest.

Ethical issues

There is none to be declared.

Abbreviations

γ	gamma
DPPH	1,1-diphenyl-2-picrylhydrazyl
PDE	phosphodiesterase
EPA	eicosapentaenoic acid
MRSA	methicillin-resistant *Staphylococcus aureus*
ACE	angiotensin-converting enzyme
ATCC	American Type Culture Collection
PUFA	polyunsaturated fatty acids
HUFA	highly unsaturated fatty acids
GLA	gamma-linolenic acid
ARA	arachidonic acid
DHA	docosahexaenoic acid
GABA	gamma-amino-butyric acid
EPS	exopolysaccharides
LAB	lactic acid bacteria
CO_2	carbon-di-oxide
HIV	Human Immunodeficiency Virus

Author details

Charu Gupta* and Dhan Prakash
Amity Institute for Herbal Research and Studies, Amity University-UP, Noida, India

*Address all correspondence to: charumicro@gmail.com

References

[1] Chapman AS, Stévant P, Larssen WE, Stévant P, Larssen WE. Food or fad? Challenges and opportunities for including seaweeds in a Nordic diet. Botanica Marina. 2015;**58**:423-433

[2] Cornish M, Critchley A, Mouritsen O. Consumption of seaweeds and the human brain. Journal of Applied Phycology. 2017;**29**:2377-2398. DOI: 10.1007/s10811-016- 1049-3

[3] Buschmann AH, Camus C, Infante J, Neori A, Israel Á, Hernández-González MC, et al. Seaweed production: Overview of the global state of exploitation, farming and emerging research activity. European Journal of Phycology. 2017;**52**(4):391-406. DOI: 10.1080/09670262.2017.1365175

[4] Elena M. Mailiohydrin, a cytotoxic chamigrene dibromohydrin from a phillippine *Laurencia* species. Journal of Natural Products. 2003;**64**:790-791

[5] Bhagavathy S, Sumathi P, Bell IJB. Green algae chlorococcumhumicola—A new source of bioactive compounds with antimicrobial activity. Asian Pacific Journal of Tropical Biomedicine. 2011;**1**:S1-S7. DOI: 10.1016/S2221-1691(11)60111-1

[6] Dewapriya P, Se-kwon K. Marine microorganisms: An emerging avenue in modern nutraceuticals and functional foods. Food Research International. 2014;**56**:115-125. DOI: 10.1016/j.foodres.2013.12.022

[7] Aneiros A, Garateix A. Bioactive peptides from marine sources: Pharmacological properties and isolation procedures. Journal of Chromatography, B: Analytical Technologies in the Biomedical and Life Sciences. 2004;**803**:41-53. DOI: 10.1016/j.jchromb.2003.11.005

[8] Tseng CK. Algal biotechnology industries and research activities in China. Journal of Applied Phycology. 2001;**13**:375-380. DOI: 10.1023/A:1017972812576

[9] Grobbelaar JU. Algal biotechnology: Real opportunities for Africa. South African Journal of Botany. 2004;**70**:140-144. DOI: 10.1016/S0254-6299(15)30274-X

[10] Volkman JK. Sterols in microorganisms. Applied Microbiology and Biotechnology. 2003;**60**:495-506. DOI: 10.1007/s00253-002-1172-8

[11] Inagaki F, Nunoura T, Nakagawa S, Teske A, Lever M, Lauer A, et al. Biogeographical distribution and diversity of microbes in methane hydrate-bearing deep marine sediments on the Pacific Ocean margin. Proceedings of the National Academy of Sciences of the United States of America. 2006;**103**:2815-2820. DOI: 10.1073/pnas.0511033103

[12] Okami YJ. The search for bioactive metabolites forms marine bacteria. Marine Biotechnology. 1993;**1**:59-65

[13] Trischman JA, Tapiolas DM, Jensen PR, Dwight R, McKee TC, Ireland CM, et al. Salinamides A and B: Anti-inflammatory depsipeptides from a marine *Streptomycete*. Journal of the American Chemical Society. 1994;**116**:757-758. DOI: 10.1021/ja00081a042

[14] Belofsky GN, Jensen PR, Renner MK, Fenical W. New cytotoxic sesquiterpenoid nitrobenzoyl esters from a marine isolate of the fungus *Aspergillus versicolor*. Tetrahedron. 1998;**54**:1715-1724. DOI: 10.1016/S0040-4020(97)10396-9

[15] Romanenko LA, Uchino M, Kalinovskaya NI, Mikhailov VV.

Isolation, phylogenetic analysis and screening of marine mollusc-associated bacteria for antimicrobial, hemolytic and surface activities. Microbiological Research. 2008;**163**:633-644. DOI: 10.1016/j.micres.2006.10.001

[16] Isnansetyo A, Kamei Y. Bioactive substances produced by marine isolates of *Pseudomonas*. Journal of Industrial Microbiology & Biotechnology. 2009;**36**:1239-1248. DOI: 10.1007/s10295-009-0611-2

[17] Romanenko LA, Uchino M, Tanaka N, Frolova GM, Slinkina NN, Mikhailov VV. Occurrence and antagonistic potential of *Stenotrophomonas* strains isolated from deep-sea invertebrates. Archives of Microbiology. 2008;**189**:337-344. DOI: 10.1007/s00203-007-0324-8

[18] Romanenko LA, Uchino M, Tebo BM, Tanaka N, Frolova GM, Mikhailov VV. *Pseudomonas marincola* sp. nov., isolated from marine environments. International Journal of Systematic and Evolutionary Microbiology. 2008;**58**:706-710. DOI: 10.1099/ijs.0.65406-0

[19] Isnansetyo A, Kamei Y. MC21-A, a bactericidal antibiotic produced by a new marine bacterium, *Pseudoalteromonas phenolica* sp. nov. O-BC30T against methicillin-resistant *Staphylococcus aureus*. Antimicrobial Agents and Chemotherapy. 2003;**47**:480-488. DOI: 10.1128/AAC.47.2.480-488.2003

[20] Bhatnagar I, Kim S-K. Immense essence of excellence: Marine microbial bioactive compounds. Marine Drugs. 2010;**8**:2673-2701. DOI: 10.3390/md8102673

[21] Franks A, Haywood P, Holmström C, Egan S, Kjelleberg S, Kumar N. Isolation and structure elucidation of a novel yellow pigment from the marine bacterium *Pseudoalteromonas tunicata*. Molecules. 2005;**10**:1286-1291. DOI: 10.3390/10101286

[22] Egan S, James S, Holmström C, Kjelleberg S. Correlation between pigmentation and antifouling compounds produced by *Pseudoalteromonas tunicata*. Environmental Microbiology. 2002;**4**:433-442. DOI: 10.1046/j.1462-2920.2002.00322.x

[23] König G, Kehraus S, Seibert S, Abdel-Lateff A, Müller D. Natural products from marine organisms and their associated microbes. ChemBioChem. 2006;**7**:229-238. DOI: 10.1002/cbic.200500087

[24] Burke C, Thomas T, Egan S, Kjelleberg S. The use of functional genomics for the identification of a gene cluster encoding for the biosynthesis of an antifungal tambjamine in the marine bacterium *Pseudoalteromonas tunicata*. Environmental Microbiology. 2007;**9**:814-818. DOI: 10.1111/j.1462-2920.2006.01177.x

[25] Bhadury P, Mohammad BT, Wright PC. The current status of natural products from marine fungi and their potential as anti-infective agents. Journal of Industrial Microbiology & Biotechnology. 2006;**33**:325-337. DOI: 10.1007/s10295-005-0070-3

[26] Newman DJ, Hill RT. New drugs from marine microbes: The tide is turning. Journal of Industrial Microbiology & Biotechnology. 2006;**33**:539-544. DOI: 10.1007/s10295-006-0115-2

[27] Kjer J, Debbab A, Aly AH, Proksch P. Methods for isolation of marine-derived endophytic fungi and their bioactive secondary products. Nature Protocols. 2010;**5**:479-490. DOI: 10.1038/nprot.2009.233

[28] Du L, Zhu T, Fang Y, Liu H, Gu Q, Zhu W. Aspergiolide A, a novel anthraquinone derivative with naphtho(1,2,3-de)chromene-2,7-dione skeleton isolated from a marine-derived fungus *Aspergillus glaucus*. Tetrahedron. 2007;**63**:1085-1088. DOI: 10.1016/j.tet.2006.11.074

[29] Du L, Feng T, Zhao B, Li D, Cai S, Zhu T, et al. Alkaloids from a deep ocean sediment-derived fungus *Penicillium* sp. and their antitumor activities. The Journal of Antibiotics. 2010;**63**:165-170. DOI: 10.1038/ja.2010.11

[30] Kwong TF, Miao L, Li X, Qian PY. Novel antifouling and antimicrobial compound from a marine-derived fungus *Ampelomyces* sp. Marine Biotechnology. 2006;**8**:634-640. DOI: 10.1007/s10126-005-6146-2

[31] Xiong H, Qi S, Xu Y, Li M, Qian P-Y. Antibiotic and antifouling compound production by the marine-derived fungus *Cladosporium* sp. F14. Journal of Hydro-Environment Research. 2009;**2**:264-270

[32] Gai Y, Zhao LL, Hu CQ, Zhang HP. Fusarielin E, a new antifungal antibiotic from *Fusarium* sp. Chinese Chemical Letters. 2007;**18**:954-956. DOI: 10.1016/j.cclet.2007.05.048

[33] Li X, Kim SK, Nam KW, Kang JS, Choi HD, Son BW. A new antibacterial dioxopiperazine alkaloid related to gliotoxin from a marine isolate of the fungus *Pseudallescheria*. Journal of Antibiotics (Tokyo). 2006;**59**:248-250. DOI: 10.1038/ja.2006.35

[34] Zhang D, Yang X, Kang JS, Choi HD, Son BW. Chlorohydroaspyrones A and B, antibacterial aspyrone derivatives from the marine-derived fungus *Exophiala* sp. Journal of Natural Products. 2008;**71**:1458-1460. DOI: 10.1021/np800107c

[35] Stadler M, Anke H, Sterner O. Metabolites with nematicidal and antimicrobial activities from the ascomycete *Lachnum papyraceum* (Karst.) Karst. III. Production of novel isocoumarin derivatives, isolation and biological activities. The Journal of Antibiotics. 1995;**48**:261-266. DOI: 10.7164/antibiotics.48.154

[36] Nenkep VN, Yun K, Li Y, Choi HD, Kang JS, Son BW. New production of haloquinones, bromochlorogentisylquinones A and B, by a halide salt from a marine isolate of the fungus *Phoma herbarum*. The Journal of Antibiotics. 2010;**63**:199-201. DOI: 10.1038/ja.2010.15

[37] Takahashi Y, Omura S. Isolation of new actinomycete strains for the screening of new bioactive compounds. The Journal of General and Applied Microbiology. 2003;**49**:141-154. DOI: 10.2323/jgam.49.141

[38] Olano C, Méndez C, Salas JA. Antitumor compounds from marine actinomycetes. Marine Drugs. 2009;**7**:210-248. DOI: 10.3390/md7020210

[39] Prudhomme J, McDaniel E, Ponts N, Bertani S, Fenical W, Jensen P, et al. Marine actinomycetes: A new source of compounds against the human malaria parasite. PLoS One. 2008;**3**:e2335. DOI: 10.1371/journal.pone.0002335

[40] El-Gendy MM, Hawas UW, Jaspars M. Novel bioactive metabolites from a marine derived bacterium *Nocardia* sp. ALAA 2000. Journal of Antibiotics (Tokyo). 2008;**61**:379-386. DOI: 10.1038/ja.2008.53

[41] McArthur KA, Mitchell SS, Tsueng G, Rheingold A, White DJ, Grodberg J, et al. Lynamicins A-E, chlorinated bisindole pyrrole antibiotics from a novel marine actinomycete. Journal of Natural Products.

2008;**71**:1732-1737. DOI: 10.1021/np800286d

[42] Cho JY, Williams PG, Kwon HC, Jensen PR, Fenical W. Lucentamycins A–D, cytotoxic peptides from the marine-derived actinomycete *Nocardiopsis lucentensis*. Journal of Natural Products. 2007;**70**:1321-1328. DOI: 10.1021/np070101b

[43] Hawas UW, Shaaban M, Shaaban KA, Speitling M, Maier A, Kelter G, et al. Cytotoxic isoquinolinequinones from a marine streptomycete. Journal of Natural Products. 2009;**72**:2120-2124. DOI: 10.1021/np900160g

[44] Pérez M, Crespo C, Schleissner C, Rodríguez P, Zúñiga P, Reyes F. Tartrolon D, a cytotoxic macrodiolide from the marine-derived actinomycete *Streptomyces* sp. MDG-04-17-069. Journal of Natural Products. 2009;**72**:2192-2194. DOI: 10.1021/np9006603

[45] Maloney KN, Macmillan JB, Kauffman CA, Jensen PR, Dipasquale AG, Rheingold AL, et al. Lodopyridone, a structurally unprecedented alkaloid from a marine actinomycete. Organic Letters. 2009;**11**:5422-5424. DOI: 10.1021/ol901997k

[46] Carlson JC, Li S, Burr DA, Sherman DH. Isolation and characterization of Tirandamycins from a marine-derived *Streptomyces* sp. Journal of Natural Products. 2009;**72**:2076-2079. DOI: 10.1021/np9005597

[47] Maskey RP, Helmke E, Kayser O, Fiebig HH, Maier A, Busche A, et al. Anti-cancer and antibacterial trioxacarcins with high anti-malaria activity from a marine *Streptomycete* and their absolute stereochemistry. Journal of Antibiotics (Tokyo). 2004;**57**:771-779. DOI: 10.7164/antibiotics.57.771

[48] Xu Y, He H, Schulz S, Liu X, Fusetani N, Xiong H, et al. Potent antifouling compounds produced by marine *Streptomyces*. Bioresource Technology. 2010;**101**:1331-1336. DOI: 10.1016/j.biortech.2009.09.046

[49] Iwata F, Sato S, Mukai T, Yamada S, Takeo J, Abe A, et al. Lorneic acids, trialkyl-substituted aromatic acids from a marine-derived actinomycete. Journal of Natural Products. 2009;**72**:2046-2048. DOI: 10.1021/np900353y

[50] Lopes VR, Fernández N, Martins RF, Vasconcelos V. Primary screening of the bioactivity of brackish water cyanobacteria: Toxicity of crude extracts to *Artemia salina* larvae and *Paracentrotus lividus* embryos. Marine Drugs. 2010;**8**:471-482. DOI: 10.3390/md8030471

[51] Luesch H, Harrigan GG, Goetz G, Horgen FD. The cyanobacterial origin of potent anticancer agents originally isolated from sea hares. Current Medicinal Chemistry. 2002;**9**:1791-1806. DOI: 10.2174/0929867023369051

[52] Sivonen K, Leikoski N, Fewer DP, Jokela J. Cyanobactins-ribosomal cyclic peptides produced by cyanobacteria. Applied Microbiology and Biotechnology. 2010;**86**:1213-1225. DOI: 10.1007/s00253-010-2482-x

[53] Simmons TL, Engene N, Ureña LD, Romero LI, Ortega-Barría E, Gerwick L, et al. Viridamides A and B, lipodepsipeptides with antiprotozoal activity from the marine cyanobacterium *Oscillatoria nigro-*Viridis. Journal of Natural Products. 2008;**71**:1544-1550. DOI: 10.1021/np800110e

[54] Allmendinger A, Spavieri J, Kaiser M, Casey R, Hingley-Wilson S, Lalvani A, et al. Antiprotozoal, anti-mycobacterial and cytotoxic potential of twenty-three British and Irish red algae. Phytotherapy Research. 2010;**24**:1099-1103. PMID: 20077438

[55] Desbois AP, Mearns-Spragg A, Smith VJ. A fatty acid from the diatom *Phaeodactylum tricornutum* is antibacterial against diverse bacteria including multi-resistant *Staphylococcus aureus* (MRSA). Marine Biotechnology. 2009;**11**:45-52. DOI: 10.1007/s10126-008-9118-5

[56] Plaza M, Cifuentes A, Ibanez E. In the search of new functional food ingredients from algae. Trends in Food Science and Technology. 2008;**19**:31-39. DOI: 10.1016/j.tifs.2007.07.012

[57] Gomez-Ordonez E, Jimenez-Escrig A, Ruperez P. Dietary fibre and physico-chemical properties of several edible seaweeds from the Northwestern Spanish coast. Food Research International. 2010;**9**:2289-2294. DOI: 10.1016/j.foodres.2010.08.005

[58] Kim EY, Kim DG, Kim YR, Hwang HJ, Nam TJ, Kong IS. An improved method of protein isolation and proteome analysis with *Saccharina japonica* (Laminariales) incubated under different pH conditions. Journal of Applied Phycology. 2011;**23**:123-130. DOI: 10.1007/s10811-010-9550-6

[59] Thomas TRA, Kavlekar DP, LokaBharathi PA. Marine drugs from sponge-microbe association—A review. Marine Drugs. 2010;**8**:1417-1468. DOI: 10.3390/md8041417

[60] Baker PW, Kennedy J, Dobson ADW, Marchesi JR. Phylogenetic diversity and antimicrobial activities of Fungi associated with *Haliclona simulans* isolated from Irish coastal waters. Marine Biotechnology. 2009;**11**:540-547. DOI: 10.1007/s10126-008-9169-7

[61] Gandhimathi R, Arunkumar M, Selvin J, Thangavelu T, Sivaramakrishnan S, Kiran GS, et al. Antimicrobial potential of sponge associated marine actinomycetes. Journal de Mycologie Médicale. 2008;**18**:16-22. DOI: 10.1016/j.mycmed.2007.11.001

[62] Selvin J, Shanmughapriya S, Gandhimathi S, Kiran GS, Ravji TR, Natarajaseenivasan K, et al. Optimization and production of novel antimicrobial agents from sponge associated marine actinomycetes *Nocardiopsis dassonvillei* MAD08. Applied Microbiology and Biotechnology. 2009;**83**:435-445. DOI: 10.1007/s00253-009-1878-y

[63] Byun HG, Zhang H, Mochizuki M, Adachi K, Shizuri Y, Lee WJ, et al. Novel antifungal diketopiperazine from marine fungus. Journal of Antibiotics (Tokyo). 2003;**56**:102-106. DOI: 10.7164/antibiotics.56.102

[64] Kalinovskaya NI, Ivanova EP, Alexeeva YV, Gorshkova NM, Kuznetsova TA, Dmitronek AD, et al. Low-molecular-weight, biologically active compounds from marine *Pseudoalteromonas* species. Current Microbiology. 2004;**48**:441-446. DOI: 10.1007/s00284-003-4240-0

[65] Radjasa OK. Marine invertebrate-associated bacteria in coral reef ecosystems as a new source of bioactive compounds. Journal of Coastal Development. 2004;**7**:65-70

[66] Zheng L, Han X, Chen H, Lin W, Yan X. Marine bacteria associated with marine macro-organisms: The potential antimicrobial resources. Annales de Microbiologie. 2005;**55**:119-124

[67] Soman AG, Gloer JB, Angawi RF, Wicklow DT, Dowd PF. Vertilecanins: New phenopicolinic acid analogs from *Verticillium lecanii*. Journal of Natural Products. 2001;**64**:189-192. DOI: 10.1021/np000094q

[68] Che Y, Gloer JB, Wicklow DT. Phomadecalins A-D and Phomapentenone A: New bioactive metabolites from *Phoma* sp. NRRL

25697, a fungal colonist of *Hypoxylon Stromata*. Journal of Natural Products. 2002;**65**:399-402. DOI: 10.1021/np010519o

[69] Raposo MFD, de Morais R, de Morais A. Health applications of bioactive compounds from marine microalgae. Life Sciences. 2013;**93**:479-486. DOI: 10.1016/j.lfs.2013.08.002

[70] Satpute SK, Banat IM, Dhakephalkar PK, Banpurkar AG, Chopade BA. Biosurfactants, bioemulsifiers and exopolysaccharides from marine microorganisms. Biotechnology Advances. 2007;**28**:436-450. DOI: 10.1016/j.biotechadv.2010.02.006

[71] Arena A, Gugliandolo C, Stassi G, Pavone B, Iannello D, Bisignano G, et al. An exopolysaccharide produced by *Geobacillus thermodenitrificans* strain B3-72: Antiviral activity on immunocompetent cells. Immunology Letters. 2009;**123**:132-137. DOI: 10.1016/j.imlet.2009.03.001

[72] Ande MP, Syamala K, Rao PS, Mohan KM, Lingam SS. Marine nutraceuticals. Aquaculture Times. 2017;**3**(2):06-09

[73] Suleria HAR, Osborne S, Masci P, Gobe G. Marine-based nutraceuticals: An innovative trend in the food and supplement industries. Marine Drugs. 2015;**13**:6336-6351. DOI: 10.3390/md13106336

[74] Das P, Mukherjee S, Sen R. Improved bioavailability and biodegradation of a model polyaromatic hydrocarbon by a biosurfactant producing bacterium of marine origin. Chemosphere. 2008;**72**:1229-1234. DOI: 10.1016/j.chemosphere.2008.05.015

[75] Das P, Mukherjee S, Sen R. Biosurfactant of marine origin exhibiting heavy metal remediation properties. Bioresource Technology. 2009;**100**:4887-4890. DOI: 10.1016/j.biortech.2009.05.028

[76] Davidson SK, Allen SW, Lim GE, Anderson CM, Haygood MG. Evidence for the biosynthesis of bryostatins by the bacterial symbiont "*Candidatus Endobugula sertula*" of the Bryozoan *Bugula neritina*. Applied and Environmental Microbiology. 2001;**67**(10):4531-4537. DOI: 10.1128/AEM.67.10.4531-4537.2001

[77] Chalova V, Lingbeck J, Kwon Y, Ricke S. Extracellular anti-mutagenic activities of selected probiotic *Bifidobacterium* and *Lactobacillus* spp. as a function of growth phase. Journal of Environmental Science and Health, Part B. 2008;**43**(2):193-198. DOI: 10.1080/03601230701795262

[78] Forsythe P, Bienenstock J. Immunomodulation by commensal and probiotic bacteria. Immunological Investigations. 2010;**39**:429-448. DOI: 10.3109/08820131003667978

[79] Je JY, Kim SK. Chitosan as potential marine nutraceutical. Advances in Food and Nutrition Research. 2012;**65**:121-135. DOI: 10.1016/B978-0-12-416003-3.00021-4

[80] Kim SK. Marine medical foods: Implications and applications: Animals and microbes. Advances in Food and Nutrition Research. 2012;**65**:280. DOI: 10.1016/B978-0-12-416003-3.00001-9

[81] Rayapu L, Makkar F, Chakraborty K, Valluru L. Phytochemical evaluation and antimicrobial activity of *Gracilaria opuntia*: An important anti-diabetic red marine macroalgae. International Journal of Current Pharmaceutical Research. 2017;**9**(6):37-41. DOI: 10.22159/ijcpr.2017v9i6.23426

[82] Rajakumar R, Allwin Prem Singh Y. Preliminary phytochemical

and antimicrobial studies on the crude extract of red algae *Gracilaria edulis* against clinical isolates. European Journal of Pharmaceutical and Medical Research. 2017;**4**(07):763-766

[83] Mehdinezhad N, Ghannadi A, Yegdaneh A. Phytochemical and biological evaluation of some *Sargassum* species from Persian Gulf. Research in Pharmaceutical Sciences. 2016;**11**(3):243-249

[84] Rajendran N, Selvan K, Sobana P, Logeswar V, Kathiresan E, Tamilselvi A, et al. Phytochemicals, antimicrobial and antioxidant screening from five different marine microalgae. Journal of Chemical and Pharmaceutical Sciences. 2014;(Special Issue 2):78-85

[85] Blunt JW, Copp BR, Munro MHG, Northcote PT, Prinsep MR. Marine natural products. Natural Product Reports. 2004;**21**:1-49. DOI: 10.1039/b305250h

[86] Mordor Intelligence Global Nutraceuticals Market—Growth, Trends and Forecasts (2015-2020). Available from: http://www.mordorintelligence.com/industry-reports/global-nutraceuticals-market-industry [Accessed: August 1, 2015]

[87] Research and Markets: Nutraceuticals—2012. Global Strategic Business Report Annual Estimates and Forecasts for 2010-2018. 2012. Available from: http://www.researchandmarkets.com/research/n54vdx/nutraceuticals [Accessed: September 20, 2012]

[88] Rovira MA, Grau M, Castañer O, Covas MI, Schröder H, REGICOR Investigators. Dietary supplement use and health-related behaviors in a Mediterranean population. Journal of Nutrition Education and Behavior. 2013;**45**(5):386-391

[89] Gupta C, Prakash D. Nutraceuticals for geriatrics. Journal of Traditional and Complementary Medicine. 2015;**5**(1):5-14